for ArcGIS 10.3.x

GISTUTORIAL

Spatial Analysis Workbook

David W. Allen

Esri Press
REDLANDS│CALIFORNIA

Esri Press, 380 New York Street, Redlands, California 92373-8100

Originally published as *GIS Tutorial II: Spatial Analysis Workbook*

20 19 18 17 16   1 2 3 4 5 6 7 8 9 10

Printed in the United States of America

Ask for Esri Press titles at your local bookstore or order by calling 800-447-9778, or shop online at esri.com/esripress. Outside the United States, contact your local Esri distributor or shop online at eurospanbookstore.com/esri.

Esri Press titles are distributed to the trade by the following:

In North America:
Ingram Publisher Services
Toll-free telephone: 800-648-3104
Toll-free fax: 800-838-1149
E-mail: customerservice@ingrampublisherservices.com

In the United Kingdom, Europe, Middle East and Africa, Asia, and Australia:
Eurospan Group
3 Henrietta Street
London WC2E 8LU
United Kingdom
Telephone: 44(0) 1767 604972
Fax: 44(0) 1767 601640
E-mail: eurospan@turpin-distribution.com

Contents

Preface

Spatial analysis involves the problem-solving aspect of GIS. From a cursory evaluation, the tools are basic—buffers, overlays, selections—but when combined in a particular sequence they can reveal things about the data that cannot be seen in a spreadsheet or chart. With this book, the reader will not create any new data, but will generate new files based on existing data. That is because analysis is not about creating new data but about making existing data say new things. The key is to know the tools well and to design the right sequences to bring the big picture in view. *GIS Tutorial 2: Spatial Analysis Workbook*, 10.3.*x* edition, focuses on data presentation and equips readers with the skills to build "big picture" maps.

This workbook is a compilation of tutorials and exercises based on the two-volume set *The Esri Guide to GIS Analysis* by Andy Mitchell. For best results, you should read the corresponding pages in *The Esri Guide to GIS Analysis* before each tutorial.

Chapters 1–6 correspond with chapters from *The Esri Guide to GIS Analysis,* volume 1. These correlations are easily identified because corresponding chapters have the same titles. Chapter 1 concentrates on symbology and categorization with the viewer in mind. The material in chapter 2 deals with mapping quantities and recognizing patterns through classification. In chapter 3, the lessons cover density mapping for value comparison. Chapter 4 focuses on creating boundaries using visual overlays for performing inside–outside analysis. Chapter 5 addresses the analysis of distance relationships between features. In chapter 6, readers learn how to represent data fluctuations over time.

Chapters 7–9 correspond to content from *The Esri Guide to GIS Analysis, volume 2*. The tutorials and exercises in chapter 7 deal with displaying geographic distribution to aid analysis. Chapters 8 and 9 address spatial statistics and introduce a higher level of mathematics for establishing a statistical confidence level for analysis.

Each tutorial contains several elements that build context and engage critical thought to reinforce the skills learned. Review sections appear after each exercise to recap the actions taken. Study questions are provided to encourage further analysis. "Your turn" tasks appear within certain tutorials to support independent demonstration of skills learned. The workbook also offers "Other real-world examples" throughout, which provide the reader with real-world scenarios that use the skills covered. Additionally, the "Independent projects" section outlines six different scenarios for independent projects that build on the skills readers acquire in the tutorials.

This workbook is geared toward a more advanced readership than the introductory *GIS Tutorial 1: Basic Workbook* and was written for those who want to learn more about analysis tools in ArcGIS and how to use them. It assumes an existing knowledge of ArcMap and ArcCatalog and requires the use of ArcGIS extensions, as well as third-party tools and scripts that are included in the student and instructor data.

Trial software and instructional data

For students: *GIS Tutorial 2* comes with exercise data and a trial of ArcGIS, which you can access in the Esri Press online resources at esri.com/esripress-resources. The trial comes with ArcGIS 10.3.1.*x* for Desktop. If you have an earlier version of ArcGIS for Desktop, you will need to uninstall it before loading the trial software.

For instructors: instructors can request instructor resources for this book at esri.com /esripress. This includes an instructor guide with outlines of each tutorial and the answers to the study questions, plus completed data and map documents for each tutorial and exercise.

Fixing broken path links in layer files

If you install the exercise maps and data to a folder structure other than the default, you must fix the broken path links for the layer (.lyr) files. To fix a broken data link, open ArcCatalog (or use the Catalog window in ArcMap), right-click the layer file, and click Properties. Click the Source tab, and the Data Source box displays the expected file location. Click the Set Data Source button, and navigate to the drive and locate where you installed the exercise data. For example, if you installed the data on drive E, your location might be something like the following (including the feature dataset name and feature class name): \\EsriPress\GIST2\Data\CityOfOleander.gdb\FireDepartment\FileName. You can move the dialog box to the side and see the correct feature data name and feature class name while you are setting it. Once you have reset the data source location, click OK to close the Properties dialog box.

The following files in the Data folder may need to be fixed:
    \Data\DFWRegionRoads.lyr
    \Data\Flood Areas.lyr
    \Data\LandUse.lyr
    \Data\LandUseCodes.lyr
    \Data\Lot Boundaries.lyr
    \Data\Site1_Time.lyr
    \Data\Site2_Time.lyr
    \Data\ZoningDistricts.lyr

Acknowledgments

As the program coordinator of the GIS degree offerings at Tarrant County College, I see a lot of what they call "nontraditional" students in my classes—students who have been in the work force for some time and are returning to college for additional education. Nothing impresses or inspires me more than when someone who decides to add college classes to their already busy schedule comes into my classroom and says, "Help me be better." It is to these students that I give the most recognition.

I'd like to thank Esri Press and Andy Mitchell for producing the two volumes that inspired this one. We've used these books in class for many years, and I developed these tutorials to reinforce the topics that are presented there. Thanks to Dr. Lauren Scott Griffin, who helped review the statistics tutorials, and good friend Clint Brown of Esri, who always makes this technology seem so approachable.

Thanks also to Doug Zedler of the City of Fort Worth Fire Department and Gini Connolly of the City of Hurst, Texas, for their help with the reviews. And finally, thanks to the City of Euless, Texas, administration, which allowed its rich GIS datasets to be used in the making of these tutorials. Although the data and processes are based on reality, all the scenarios are fictional and should not be associated with the City of Euless.

1

Mapping where things are

The simplest form of analysis is to show features on a map and let the viewer do the analysis in their mind's eye. It falls on the cartographer to use various colors and symbols and to group the data in a logical manner so that the viewer can clearly see the information being highlighted. More complex methods involve categorizing the data and designing symbology for each category.

Tutorial 1-1

Working with categories

The most basic of maps simply show where things are without complicated analysis. You can enhance these useful maps by symbolizing different categories. By symbolizing categories, you can show both the location and some characteristics of the features.

Learning objectives

- *Work with unique value categories*
- *Determine a display strategy*
- *Add a legend*
- *Set legend parameters*
- *Use visual analysis to see geographic patterns*

Preparation

- *Read through page 29 in Andy Mitchell's* The Esri Guide to GIS Analysis, *volume 1 (Esri Press, 2001).*

Introduction

The first and simplest type of geographic analysis is visual—that is, view it. You can display data on a map using various colors and symbology that enable the viewer to begin to see geographic patterns. But deciding what aspects of the features to highlight can take some thought.

It may be as important to map where things are not located as it is to map where they are. Visual analysis allows you to see the groupings of features, as well as areas in which features are not grouped. All of this analysis takes place in the viewer's mind, considering that visual analysis does not quantify the results. In other words, display the data using good cartographic principles, and the viewer will determine what, if any, geographic patterns might exist. Viewers do not get an explicit answer. Rather, they can determine one on their own.

You can aid the process by determining the best way to display the data, but the proper method will depend on the audience. If the audience is unfamiliar with the type of data being shown or the area of interest on the map, you may want to include more reference information about the data. You may also want to simplify the way the data is represented or use a subset of the data to make the information more easily understood by a novice audience.

Conversely, if the audience is technically savvy and familiar with the data, you can make the symbology more detailed and specific.

In making the map, you will decide what features to display and how to symbolize them. Sometimes simply using the same symbol to show where features are located is enough. For example, seeing where all the stores are in an area might give you an idea of where the shopping district is. All you really want to know is where the stores are, and not what each type of store is.

A more complex method of displaying the data is to use categories, or symbolize each feature by an attribute value in the data. This method also requires a more complex dataset. Your dataset will need a field to store a value that describes the feature's type or category. A dataset of store locations may also have a field storing the type of store: clothing, convenience, auto repair, supermarket, fast food, and so on.

For other instances, you might want to see only a subset of the data. These cases can still be shown with a single symbol but using only one value of a field. It might be fine to see where all crimes are occurring, but the auto theft task force may want to see only the places where cars have been stolen. The extra data will only confuse the map's message and possibly obscure a geographic pattern.

When making the map, you can use this "type" or "category" field to assign a different symbol to each value. You might symbolize clothing stores with a picture of a clothes hanger or a supermarket with a picture of a shopping cart. This visualization still does not involve any geographic processing of the data; you are merely showing the feature's location and symbolizing it by a type. The analysis is still taking place in the viewer's mind.

Scenario You are the GIS manager of a city of 60,000 in Texas, and the city planner is asking for a map showing zoning. The target audience is the city council, and each member is very familiar with the zoning categories and the types of projects that may be built in each area. Council members will frequently refer to this map to see in which zoning category a proposed project may fall and what effect that project might have on adjacent property. For example, a proposed concrete mixing plant would only be allowed in an industrial district and would adversely affect residential property if the plant was allowed to be near housing.

Because you will have a technically savvy audience, you can use a lot of categories and not worry about the map being too confusing or hard to read. The city planner asks that you use colors that correspond to a standard convention used for zoning. Later, you will work with setting categories and symbology.

If you have not already downloaded the data for this book, go to the Esri Press online resources at esri.com/esripress-resources. From there, you can download the exercise data, as well as a free trial of ArcGIS.

Data The first layer is a zoning dataset that contains polygons representing every zoning case ever heard by the city council. An existing field carries a code that represents the zoning category assigned to each area. This code was already set as the "value field" to define the symbology, so you will only deal with how the categories are shown. The city created this data, and a list of the zoning codes and what they mean, called a *data dictionary*, is provided later in the chapter. You may also find similar datasets from other sources that have zoning classifications, and it is important to also get their data dictionaries.

Start ArcMap and open a map document

1 On the Windows taskbar, click Start > All Programs > ArcGIS > ArcMap 10.3.*x*.

Depending on how ArcGIS and ArcMap have been installed, or which Windows operating system you are using, there may be a slightly different navigation menu to open ArcMap. You may also have an ArcMap icon on your desktop, which will start ArcMap when you double-click it.

2 In the resulting ArcMap window, click the Existing Maps header and then click "Browse for more."

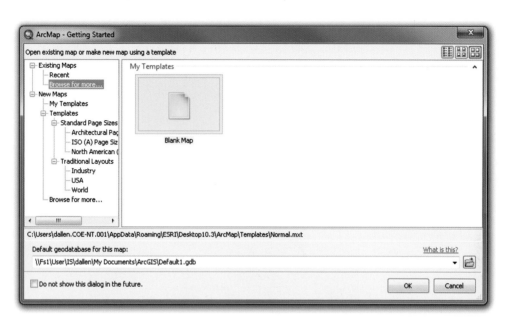

3 Browse to the drive on which the tutorial data has been installed (for example, C:\EsriPress\GIST2\Maps), click Tutorial 1-1.mxd, and then click Open.

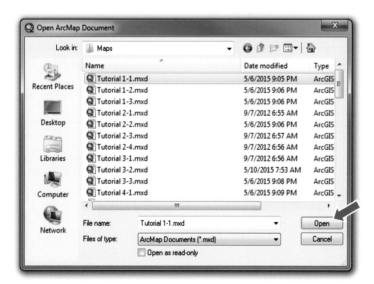

This action adds the map document to your Existing Maps list. The list includes your most commonly used maps.

The data shown is zoning categories, each zone in a different color. Note that your toolbar locations may differ from the locations shown. The city planner, for whom you are doing this work, uses this map to visually determine the zoning classification of property.

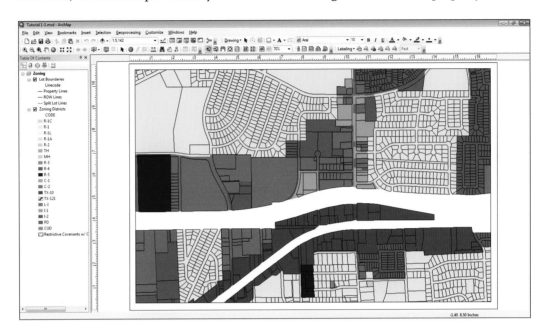

Open a layer attribute table

1 In the table of contents, right-click the Zoning Districts layer and click Open Attribute Table.

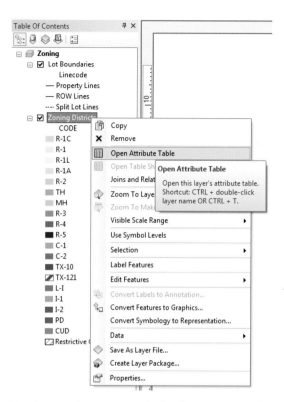

The data is shown in a tabular format. The columns are the fields in which the data is stored for this layer, and the rows represent each individual feature in the layer.

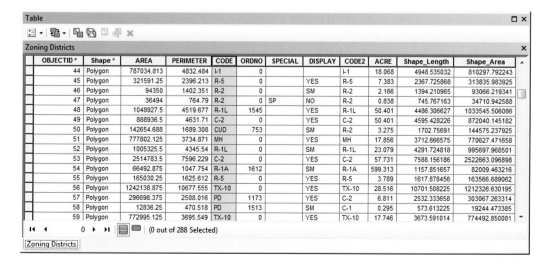

OBJECTID *	Shape *	AREA	PERIMETER	CODE	ORDNO	SPECIAL	DISPLAY	CODE2	ACRE	Shape_Length	Shape_Area
44	Polygon	787034.813	4832.484	I-1	0			I-1	18.068	4948.535032	810297.792243
45	Polygon	321591.25	2396.213	R-5	0		YES	R-5	7.383	2367.725868	313835.983925
46	Polygon	94350	1402.351	R-2	0		SM	R-2	2.166	1394.210965	93066.219341
47	Polygon	36494	764.79	R-2	0	SP	NO	R-2	0.838	745.767163	34710.942588
48	Polygon	1048927.5	4519.677	R-1L	1545		YES	R-1L	50.401	4486.306627	1033545.506086
49	Polygon	888936.5	4631.71	C-2	0		YES	C-2	50.401	4595.428226	872040.145182
50	Polygon	142654.688	1689.308	CUD	753		SM	R-2	3.275	1702.75691	144575.237925
51	Polygon	777802.125	3734.871	MH	0		YES	MH	17.856	3712.666575	770627.471658
52	Polygon	1005325.5	4345.54	R-1L	0		SM	R-1L	23.079	4291.724818	995697.968501
53	Polygon	2514783.5	7596.229	C-2	0		YES	C-2	57.731	7588.156186	2522663.096898
54	Polygon	66492.875	1047.754	R-1A	1612		SM	R-1A	599.313	1157.851657	82009.463216
55	Polygon	165030.25	1625.612	R-5	0		YES	R-5	3.789	1617.878456	163566.689062
56	Polygon	1242138.875	10677.555	TX-10	0		YES	TX-10	28.516	10701.508225	1212326.630195
57	Polygon	296698.292	2508.016	PD	1173		YES	C-2	6.811	2532.333658	303067.263314
58	Polygon	12836.25	470.518	PD	1513		SM	C-1	0.295	573.613225	19244.473385
59	Polygon	772995.125	3695.549	TX-10	0		YES	TX-10	17.746	3673.591014	774492.850081

The field CODE contains the zoning category for each parcel, representing various single-family, multifamily, commercial, and industrial uses. Other fields contain the ordinance number (ORDNO) for the zoning case, a display marker, and the acreage.

2   Close the attribute table.

Change a category label

The layer came with the code values displayed for each category. Although these values may be fine for someone who works with these codes every day and recognizes the underlying meaning, someone who is not familiar with the codes will want a more descriptive label. You can change the labels in the symbol properties of the layer.

1   In the table of contents, right-click the Zoning Districts layer and click Properties.

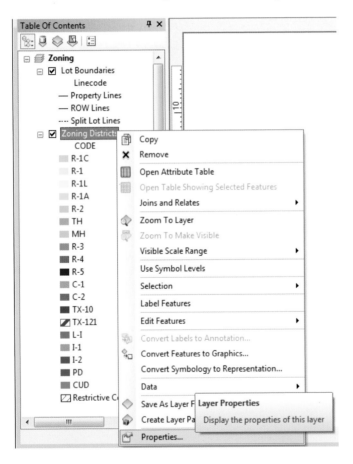

2 Click the Symbology tab.

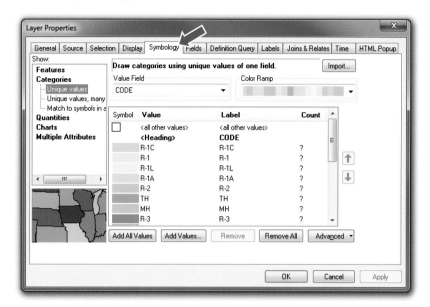

Note that the Categories selection is set to Unique values, and the Value field is set to CODE. This setting means that each unique value of the field CODE will be represented in the legend.

To a city planner or someone who is experienced with zoning districts, these codes tell the whole story. But for the layperson, these codes can be difficult to interpret. You will change the codes to match the simplified list provided by the city planner.

3 Click the R-1C entry under the Label column and type a new description: **Single Family Custom Dwellings**.

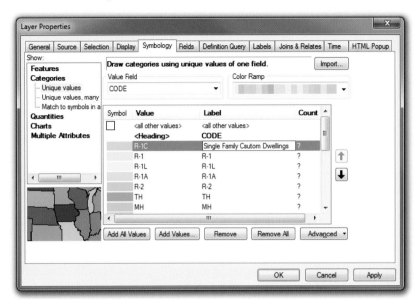

4 Click OK to close the dialog box.

The text you typed in the Label column is what will be shown in the legend.

YOUR TURN

Change all the labels to match the following list. **Note:** you can change the width of the column headers Value and Label by dragging the vertical dividers side to side to reveal the entire category description. When all the descriptions are entered, click Apply, but do not close the Layer Properties dialog box.

R-1	Single Family Detached Dwellings
R-1L	Single Family Limited Dwellings
R-1A	Single Family Attached Dwellings
R-2	Duplex Dwellings
TH	Townhouses
MH	Mobile Homes
R-3	Multi-Family Dwellings 16 units/acre
R-4	Multi-Family Dwellings 20 units/acre
R-5	Multi-Family Dwellings 25 units/acre
C-1	Neighborhood Business District
C-2	Community Business District
TX-10	Texas Hwy 10 Business District
TX-121	Texas Hwy 121 Business District
L-I	Limited Industrial District
I-1	Light Industrial District
I-2	Heavy Industrial District
PD	Planned Development
CUD	Community Unit Development

All the descriptions are in, but they are not in the order that the city planner wants. These labels can be moved to appear in the desired order.

5 Highlight the value R-1A and click the Up arrow to move it above R-1C in the list.

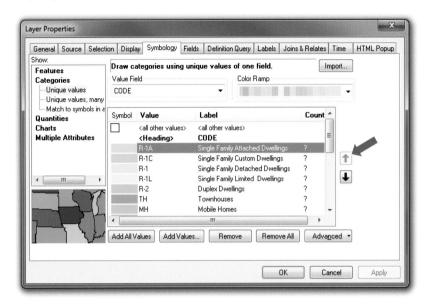

6 Click OK to close the Layer Properties dialog box.

Once you have moved the values in the correct order, your table of contents should look like the graphic.

This zoning layer is used a lot in city maps, so it is a good idea to save all this work for future use. You can save it by creating a layer file (layer file names end in ".lyr"). A layer file saves all the symbology settings for a layer, but it does not save any of the data. A layer file is useful when a dataset such as zoning must be symbolized in several different ways for different audiences, but you do not want to save the dataset several times. If you had to make updates in several files, it would lead to a problem in updating the data and keeping all the files current. With a layer file, you maintain one dataset along with several ways to symbolize the dataset.

7 Right-click the Zoning Districts layer and click Save As Layer File. Save the file as **DetailedZoning.lyr** in your \GIST2\MyExercises folder.

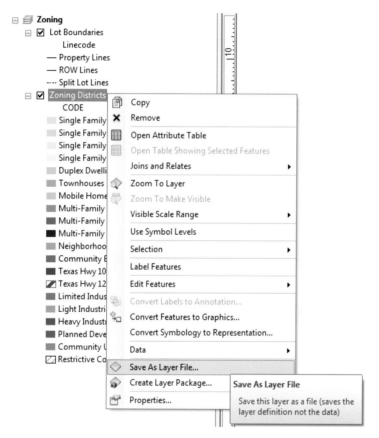

The next time you want to use the zoning data, do not add the feature class or shapefile; add the layer file, and the data will be added and already symbolized.

Add a legend to your map layout

Now that you have set everything, you can add a legend to the map layout. In fact, even if everything is not exactly the way you want or if the city planner changes his mind later, any changes you make to the table of contents will automatically be reflected in the legend.

1 On the main menu, click the Insert menu and then click Legend.

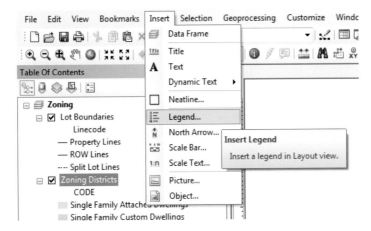

2 In the Legend Wizard window, highlight the Lot Boundaries layer in the Legend Items column and click the < button.

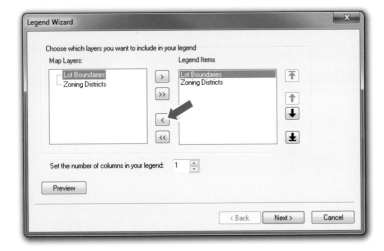

The Lot Boundaries layer is removed from the legend.

3 Set the number of columns in the legend to 2 and click Next.

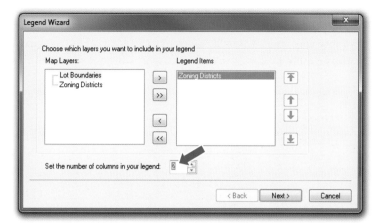

4 Change the text Legend Title to **City of Oleander**. Click the Color drop-down arrow and change the font color to Lapis Lazuli. Click Next.

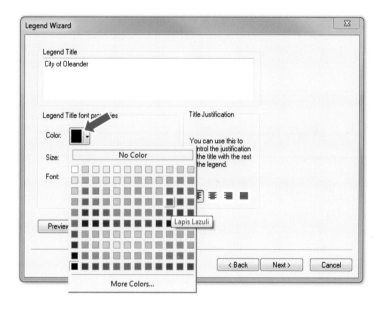

5 Click the Background color box and select Grey 10% from the drop-down list. Click Next. Complete the legend with the remaining default values by clicking Next and Finish.

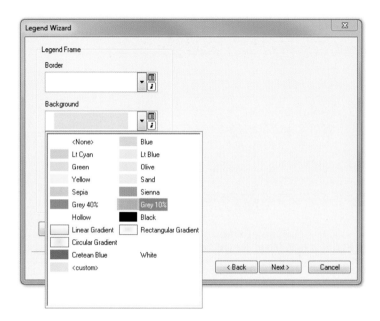

The legend is complete and is added to the map. However, it is slightly too large and not in the right place. You will use the Select Elements tool ![pointer] to resize and move it.

When a graphic element is selected, cyan boxes appear at the corners of the element. You can change the element by clicking these boxes using the Select Elements tool. Your map contains a specific spot for the legend, so you are going to use page coordinates and a set frame size in the legend's properties dialog box to place the legend.

6 With the Select Elements tool active, right-click the legend and click Properties. Click the Size and Position tab.

7 Enter the new x- and y-values for Position and the new Width and Height values for Size, as shown in the graphic. Click Apply and then click OK. You may need to click the Zoom Whole Page button ⬚ on the Layout toolbar to see the entire map.

Your completed map should look like the one shown. This type of map, showing "where things are," allows viewers to recognize and analyze items in their mind's eye. You are not identifying every feature on the map; you are only giving users a color reference that they can use to identify features. Likewise, you are not pointing out any spatial relationships in the data, but letting viewers perform visual analysis to draw their own conclusions.

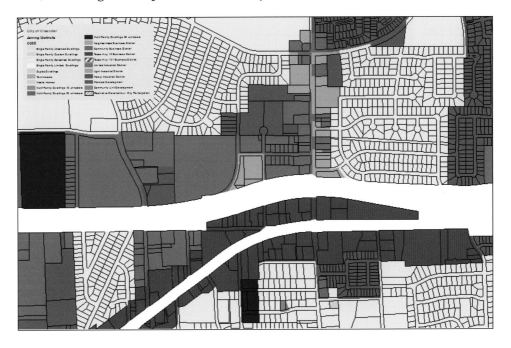

8 Save your map document as **Tutorial 1-1.mxd** to the \GIST2\MyExercises folder. If you are not continuing to the exercise, exit ArcMap.

Exercise 1-1

The first tutorial showed how to use categories to help the map viewer determine the zoning for a piece of property. The user looks at the property on the map, notes the color, and then matches that color in the legend to determine the associated zoning code. You also helped the viewer better understand zoning codes by providing a text translation of the otherwise cryptic values.

In this exercise, you will repeat the process using land-use codes, which show how property is being used, as opposed to zoning codes, which show the prescribed use of land. Land-use codes are also cryptic, so you will change the labels to a simple description. You will also add the legend to the map layout, and then resize and move the legend to an appropriate place on the map.

- Continue with the map document you created in this tutorial, or open Tutorial 1-1. mxd from the \GIST2\Maps folder.
- Turn off the Zoning Districts layer.
- Add the layer file LandUseCodes.lyr from the \GIST2\Data folder. **Note:** if you loaded the tutorial data to a location other than the default, you must repair the data path for layer files before adding them to your map. See the instructions in the preface on how to fix broken path links.
- Change the labels for the Land Use values to match the following labels:
  - A1   Single Family Detached
  - A2   Mobile Homes
  - A3   Condominiums
  - A4   Townhouses
  - A5   Single Family Detached Limited
  - AFAC   Airport Facilities
  - APR   Airport Private Land
  - AROW   Airport ROW
  - B1   Multi-Family
  - B2   Duplex

- B3　　　Triplex
- B4　　　Quadruplex
- B5　　　High Density Multi-Family
- CITY　　City Property
- CITYV　Vacant City Property
- CITYW　City Water Utilities Property
- CRH　　Church
- ESMT　Easement
- F1　　　Commercial
- F2　　　Industrial
- GOV　　Government (State or Federal)
- POS　　Public Open Space
- PRK　　Park
- PROW　Private Right-of-way
- ROW　　Right-of-way
- SCH　　School
- UNK　　Unclassified
- UTIL　　Utilities
- VAC　　Vacant

- Make any necessary changes to the legend so that it displays all the values.
- Create a layer file for the new symbology.
- Save the results as **Exercise 1-1.mxd** to the \GIST2\MyExercises folder.

What to turn in

If you are working in a classroom setting with an instructor, you may be required to submit the maps you created in tutorial 1-1.

Turn in a printed map or screen capture of the following:

> **Tutorial 1-1.mxd**
> **Exercise 1-1.mxd**

Tutorial 1-1 review

You determined that your audience was very familiar with the data being used and decided to make a detailed map of zoning and land use. Next, you checked the attribute table to see which field might contain zoning or land-use data. Since the field existed, you were able to use it to set the symbology.

The codes were rather cryptic, even for your expert audience, so you changed the display labels on the Symbology tab to include both the code and a text description of each category. This notation will help remind the viewer what each code represents.

Finally, you added a legend to the map to display the zoning or land-use codes and make the map complete.

Study questions

1. Select a piece of property on the map. Can you determine the adjacent zoning category?

2. The city council has been asked to approve a dry-cleaning plant in the Highway 10 business district, but members want to make sure that it is near other commercial uses and not residential. Can you find a place where this district has residential zoning next to it? How about commercial zoning next to it?

Other real-world examples

Each street in a centerline feature class may have a field with a code that represents its status in the thoroughfare plan. You can symbolize the streets with different line thicknesses according to the codes: a thick line for Freeways, a medium line for Major Collectors, and a thin line for Residential Streets. A quick look at the map will allow you to determine a fast route based on the amount of traffic a street is designed to handle. You may also use speed limits and get a similar result.

A forester may symbolize harvestable tree stands by their age. A darker symbol might be stands that are ready for harvesting, while a lighter color may be immature trees. A quick view of the map may show where to concentrate logging efforts to minimize the movement of equipment and impact on surrounding areas.

A dataset of all the counties in the United States may have a field that represents the political leaning of each county, whether Democrat, Republican, Green Party, Independent, and so on. A candidate for national office might symbolize the map according to these categories to determine friendly areas for fund-raising, and unfriendly areas for political conversions.

Tutorial 1-2

Controlling which values are displayed

In tutorial 1-1, you worked with all the values of a given category field. These values were specific codes that the experts use, but sometimes either for clarity or ease of use in an analysis project, you will want to display only some of the categories.

Learning objectives

- *Group symbology by category*
- *Arrange columns in a legend*
- *Hide categories*

Preparation

- *Read pages 30–36 in* The Esri Guide to GIS Analysis, *volume 1.*

Introduction

In tutorial 1-1, your audience was a technically savvy group that was very familiar with the data. Displaying a long list of zoning or land-use codes was appropriate for them because it allowed them to do the detailed visual analysis required. This scenario is not always the case, and often it may be necessary to simplify the display of the data to fit the audience.

There are two ways to simplify the data. One way is to use one of the generalization tools to physically change, or dissolve, the data into a simpler structure by removing vertices in lines and polygons. This method creates a copy of the data in a more simplified form, but it also adds to your data-maintenance chores. You will have to make any edits of the data to both datasets; or perhaps you can delete the simplified data, do your edits, and re-create the simplified dataset.

Another method to simplify the data is to control the way it is displayed. By grouping values together, you can create the same look as a dissolved dataset without creating a copy of your dataset. This method works well for visual analysis because all analysis takes place in the viewer's mind based on what they see. For more complex analysis that involves quantifying the results, the dissolve methods are preferred because extra processing may be done to the dissolved dataset that cannot be done to the entire dataset.

Scenario   The maps you made in tutorial 1-1 were a success with the city council and the city planner. They have caught the eye of the development director, who will be traveling to a developers conference to promote the city. He wants to show the zoning map to potential developers but wants the data simplified to something easier to grasp in a quick overview of the map. Developers will quickly scan the map looking for the type of zoning that is favorable to their business, and they will only stop to talk details if they see what they like. You will produce the zoning map again, but this time using general categories.

Data   You will use the same datasets as in tutorial 1-1. One change has been made in how the data is symbolized. In the previous tutorial, you used a code that included two special zoning districts that demonstrate an "a la carte" zoning style. This zoning style allows developers to define housing density and other special features of their projects to create a custom zoning category before obtaining city council approval. For this tutorial, you will use the "translated zoning," which will generalize the special zoning districts into the closest categorical zoning classifications. The translated zoning is stored in the field CODE2.

Change category labels

1   In ArcMap, open Tutorial 1-2.mxd from the location in which the tutorial data has been installed (for example, C:\EsriPress\GIST2\Maps).

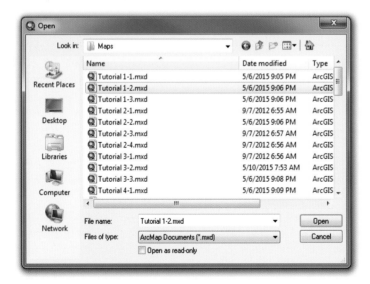

The map looks similar to the one you finished in tutorial 1-1.

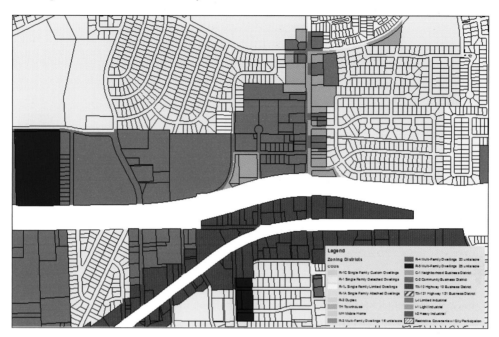

Now you will revise the map for the development director and group all the zoning codes into four categories: Single Family Residential, Multi-Family Residential, Commercial, and Industrial. These groupings will make the map suitable for a general audience that can get an overview of the city's zoning at a glance. Rather than using a dissolve function, you will change the symbology of the layer to match the groupings. It is important to note that you are not changing the underlying data in any way, but only changing how it is symbolized.

2 In the table of contents, right-click the Zoning Districts layer and click Properties.

3   Click the Symbology tab.

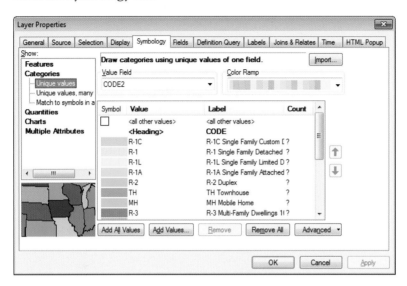

As before, all the values of the field CODE are displayed and symbolized. You want to create a simple category called "Single Family." This category will include these codes: R-1C, R-1, R-1L, R-1A, R-2, TH, and MH.

4   Click R-1C. Then press and hold Shift, and click MH. All the single-family zoning categories are highlighted. Right-click anywhere in the highlighted categories and click Group Values.

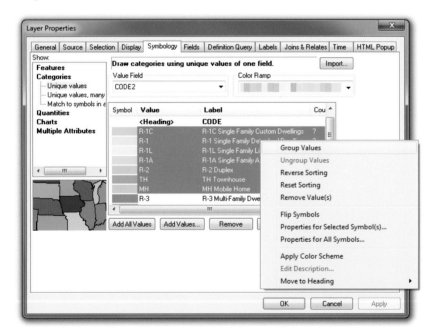

All the selected values are now shown with a single symbol. The new label strings together all the labels of the grouped values, which is not desirable. Next, you will change the label to a simple description.

5 Click the label for the grouped values and type **Single Family Residential**.

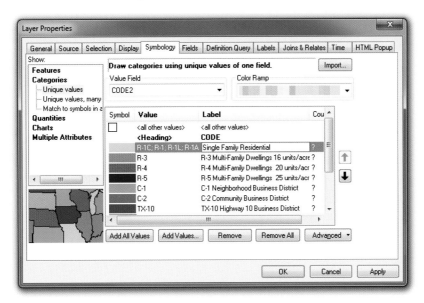

6 Click Apply and then click OK to close the dialog box.

By default, the color of the first value in the group is used for the combined values. Because single-family residential is typically shown in yellow, you can leave this color alone.

YOUR TURN

As you did with the Single Family Residential, create group symbols for the following code values:

- Groups R-3, R-4, and R-5. Change the label to **Multi-Family Residential**.
- Groups C-1, C-2, TX-10, and TX-121. Change the label to **Commercial**.
- Groups L-I, I-1, and I-2. Change the label to **Industrial**.

Click OK to accept the changes and close the dialog box.

When you are finished, the table of contents should look like the graphic.

Next, you will create a layer file of the new symbology for later use.

7 Right-click the Zoning Districts layer and click Save As Layer File. Save it to the GIST2 \MyExercises folder and name it **Grouped Zoning.lyr**.

The legend reflects the changes, but it looks a little strange. You set the legend to display two columns when there were a lot of values to show, but the simplified legend does not require two columns.

Modify the legend

1 In the layout, right-click the legend and click Properties.

2 Go to the Items tab and change the number of columns to 1. Click OK.

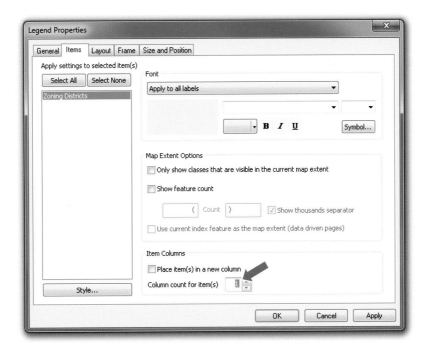

The legend displays the grouped symbology along with your updated labels in a single column. The legend should now be simple enough for the novice audience to understand.

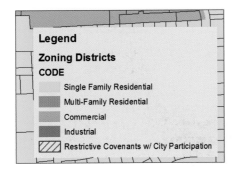

3 Look at the map, which now looks different from the more detailed version earlier. If you want to zoom into the layout so you can see the legend better, use the Zoom In button ⊕ on the Layout toolbar. Use the Zoom Whole Page button on the Layout toolbar to return the map to the original view.

4 Save your map document as **Tutorial 1-2.mxd** to the \GIST2\MyExercises folder. If you are not continuing to the exercise, exit ArcMap.

Exercise 1-2

The tutorial demonstrated how to simplify the look of your map by grouping symbol values.

In this exercise, you will repeat the process using land-use codes. These codes are also cryptic, so you will change the labels to a simple description. You will also add the legend to the map layout, and then resize and move the legend to an appropriate place on the map.

Continue with the map document you created in this tutorial, or open Tutorial 1-2. mxd from the \GIST2\Maps folder.

- Turn off the Zoning Districts layer.
- Add the layer file LandUseCodes.lyr from the \GIST2\Data folder.
- Group some of the land-use values together and update the labels to match the following definitions. For values that are not next to each other in the list, select the first one and press and hold Ctrl while selecting the others.

  - A1, A2, A3, A4, A5   Single Family Residential
  - B1, B2, B3, B4, B5   Multi-Family Residential
  - F1   Commercial
  - ESMT, F2, UNK, UTIL   Industrial
  - POS, PRK   Parks
  - CITY, CITYV, CITYW, GOV   Government
  - AROW, PROW, ROW   Right-of-way
  - AFAC, APR   Oleander International Airport
  - CRH   Church
  - SCH   School
  - VAC   Vacant

- Make any necessary changes to the legend so that it displays all the values.
- Create a layer file from the new symbology.
- Save the results as **Exercise 1-2.mxd** to the \GIST2\MyExercises folder.

What to turn in

If you are working in a classroom setting with an instructor, you may be required to submit the maps you created in tutorial 1-2.

Turn in a printed map or screen capture of the following:

> **Tutorial 1-2.mxd**
> **Exercise 1-2.mxd**

Tutorial 1-2 review

This time, your intended audience was not familiar with all the nuances of zoning and land use, so you simplified the display of the map. If more analysis steps are required, you can simplify the data. However, because you are only doing visual analysis, grouping the symbology will produce the required map without altering the data.

The simplified legend looks cleaner and clearer than the more complex legend, and you changed how it was displayed. A simple legend with five values is suitable for your audience.

Study questions

1. Look at the map and count how many different zoning categories you can identify. How complex a question can you answer with this level of data?

2. A developer is looking for a commercial corridor in which to locate a business. Can you get a general idea of where the developer should start looking?

3. The city council is reviewing an ordinance to require special fencing between light-industrial and heavy-industrial zones. Is this map suitable for that purpose? Why or why not?

Other real-world examples

Police department data may include up to eight categories of burglaries. Although this amount is great for a detailed study, the map produced for the general public might group all of the burglaries into one category.

Dallas Area Rapid Transit produces detailed maps for each bus route, but when producing a systemwide map, it groups all bus routes into one symbol. Similarly, it groups all train routes into one symbol and all light-rail lines into one symbol. This simplified map reads well on a regional scale, although it may answer only general transit questions and not guide you to a specific bus or train.

Tutorial 1-3

Limiting values to display

Legends are the key to how the map creator wants the viewer to interpret it. Sometimes all the data is shown in a complex form, and sometimes the data is simplified but still shows the entire dataset. It may also be necessary to display only a subset of the data so that the map conveys the desired message.

Learning objectives

- *Add only specific values to the legend*
- *Work with legend properties*
- *Work with definition queries*

Introduction

You have seen how to group many values together to simplify the way data is displayed on a map, but there may be times when you want to keep certain groups of values from ever showing at all. Perhaps you want to show only auto-related crime and disregard all other types of crime, or maybe show only major roads and not show local and residential streets. In this tutorial, you will look at ways to restrict which data is symbolized on your maps.

Scenario   The city planner has looked at the maps you produced earlier and was impressed. Now he has another idea for more maps that you can work on. He wants you to take the zoning map but display only the single-family zoning categories to produce an inventory of residential zoning. In this case, you must remove all other types of zoning from the map and display only the requested values.

Data   You will use the same datasets as in tutorial 1-1, but this time the datasets will not have a symbology schema already set. You will set the symbology value field to CODE and manually define what values to show.

Set symbol values

1 In ArcMap, open Tutorial 1-3.mxd from the \GIST2\Maps folder.

This time, the map has no symbology set. You will set up controls that will display only the data that meets the request that has been made.

The residential zoning inventory requires that you display only the following values: R-1, R-1A, R-1L, R-2, TH, and MH.

These values represent the single-family zoning types. The first method you will use will add only these values to the symbology editor.

2 In the table of contents, right-click the Zoning Districts layer and click Properties.

3 Click the Symbology tab. In the Show box, click Categories and then click Unique values.

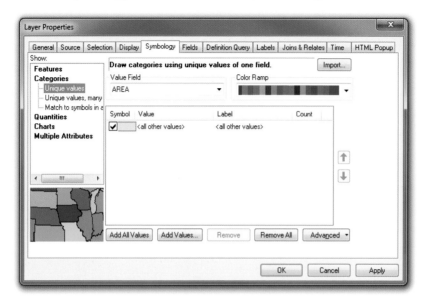

You will set the Value field to the attribute field that contains the zoning classifications. You know from tutorial 1-1 that the field is the CODE field.

4 Click the Value Field drop-down list and click CODE.

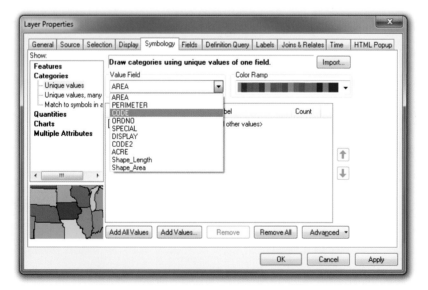

5   Click the Add Values button.

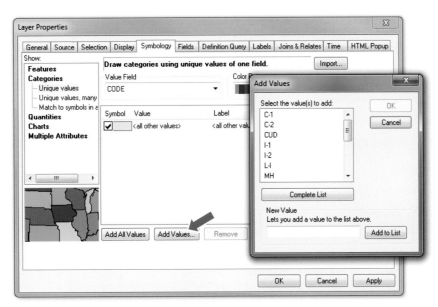

Note: the Add All Values button adds each unique occurrence of the Value field to the legend, while the Add Values button allows you to select specific field values.

6   From the list, highlight the desired single-family residential codes (listed in step 1). Press and hold Ctrl to select multiple values. If you do not see a value, click the Complete List button to refresh the list. When you have selected the right codes, click OK.

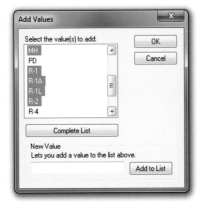

The symbology editor now shows only the selected values. If you missed one, click Add Values again and select it.

7 As in the previous tutorials, change the labels to a simple description of the zoning categories. Also, clear the <all other values> option to prevent the values that were not selected from being displayed.

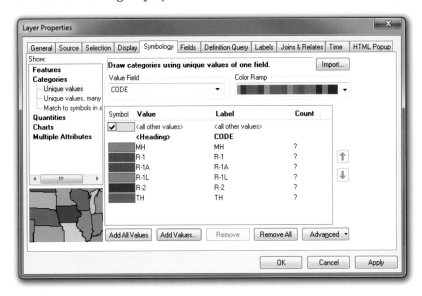

Next, you will deal with the color choices.

8 Double-click the color symbol to the left of the MH value. The Symbol Selector opens. Click Yellow and then click OK.

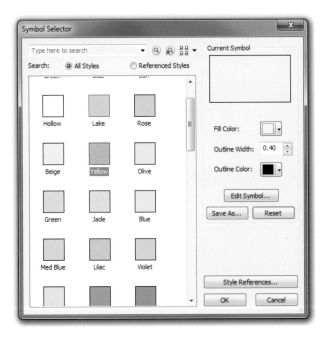

9 Repeat the process and set the other colors as follows:

- R-1   Medium Yellow
- R-1A   Tan
- R-1L   Beige
- R-2   Orange
- TH   Lt Orange

10 Verify that your dialog box matches the graphic and then click OK.

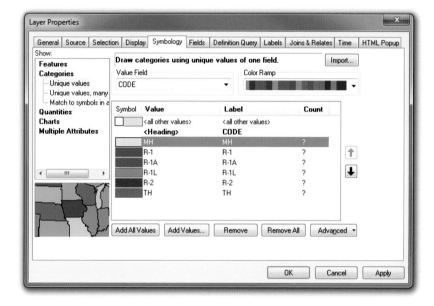

The resulting map shows only the specified residential zoning categories in the colors you chose.

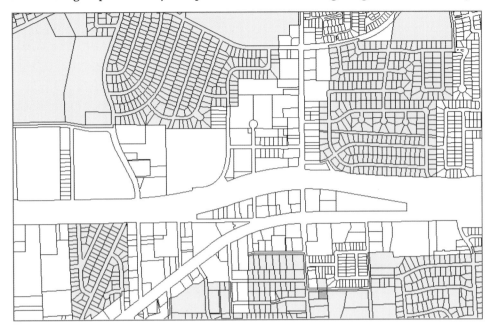

11 Create a layer file of the new symbology. Right-click the Zoning Districts layer and click Save As Layer File. Save it to the \GIST2\MyExercises folder and name it **ResidentialZoning.lyr**.

YOUR TURN

Add a single-column legend to this map displaying the values you selected and symbolized. Change the legend title to **City of Oleander**. Two things will be different from legends you have worked with in previous tutorials:

1. The legend background will be transparent, or colorless. Open the legend properties, go to the Frame tab, and set the background to 10% Grey.

2. The gap between the text and the legend box is set to a default value of 10 points. Decrease this gap by setting the background gap for X and Y to 4 points.

Many of the legend parameters are user-definable. Browse the legend properties box and experiment with a few.

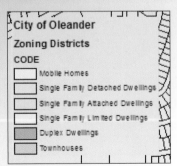

You were able to display only the categories the city planner wanted and learned a little about legends in the process. But there is one problem with this method. The city planner may want you to do some calculations or analysis against the data, and your numbers will reflect the entire dataset, not only the displayed items. Values that are not displayed are still in the database and will still be used in calculations. You will look at the attribute table to better understand the problem.

Examine the attribute table

1 In the table of contents, right-click the Zoning Districts layer and click Open Attribute Table.

The attribute table is displayed, and you can see that the CODE column contains values that you do not want shown on the map.

OBJECTID *	Shape *	AREA	PERIMETER	CODE	ORDNO	SPECIAL	DISPLAY	CODE2	ACRE	Shape_Length	Shape_Area
1	Polygon	22079.25	600.234	TX-10	0		NO	TX-10	0.507	581.697207	21117.150816
2	Polygon	139981.938	1492.252	PD	1648		YES	I-2	0	1472.455671	135636.408874
3	Polygon	394050.563	2878.861	PD	1014		SM	I-1	9.046	2844.533959	382993.567038
4	Polygon	1055427.5	4192.619	I-2	0		YES	I-2	24.229	4236.170089	1075114.505069
5	Polygon	117241.625	1370.84	I-1	0	SP	SM	I-1	2.691	1365.771874	116283.064106
6	Polygon	1728689.5	6940.227	I-1	0		YES	I-1	39.685	7030.636052	1734090.98521
7	Polygon	394069.188	3409.292	C-2	0		YES	C-2	9.047	3412.871577	390619.026774
8	Polygon	214473.063	1861.301	R-4	0		YES	R-4	4.923	1849.774471	211739.201546
9	Polygon	305606.438	2669.028	L-I	0			L-I	7.016	2700.771405	307691.01323
10	Polygon	283946.938	4168.954	PD	379		NO	R-1A	6.519	4218.482896	287151.532414
11	Polygon	33434.563	956.73	TH	379		NO	TH	0.768	1009.753944	36508.923669
12	Polygon	26335	692.575	C-2	0		SM	C-2	0.605	691.937514	26531.757215
13	Polygon	172593.688	2128.938	R-2	0		SM	R-2	3.962	2130.549204	172256.978087
14	Polygon	64251.25	1025.05	C-2	1613		SM	C-2	2.246	1045.124789	67637.241277
15	Polygon	33585.063	746.266	PD	1666		SM	R-1L	2.246	763.506672	34970.947867
16	Polygon	6921839.5	14283.724	R-1	0		LG	POS	158.904	14287.508289	6893578.324051

I◄ ◄   0   ► ►I   (0 out of 288 Selected)

Zoning Districts

2 Close the Zoning Districts attribute table when you are finished examining it.

You need a way to temporarily ignore all values except the single-family residential codes. Using a definition query is an easy way to do it. Build a query to select only the values you want, and ArcMap will ignore all the other values. It is important to note that the values will not be deleted, only ignored. Removing the definition query will restore all the dataset values.

Build a definition query

1 In the table of contents, right-click the Zoning Districts layer and click Properties.

2 In the Layer Properties dialog box, click the Definition Query tab and click Query Builder.

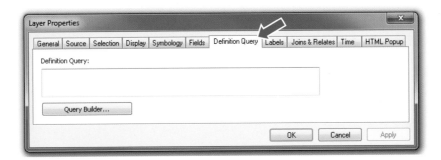

You will build a query to define the values you want to see in your map. The format for the query statement will use simple Structured Query Language (SQL) statements composed of a field name from the table; an operator, such as "greater than" (>) or "equals" (=), and a value that will be used to select fields. You can connect statements using AND or OR to select multiple values. When you use AND, the statements on both sides of the connector must be true for a feature to be selected. If you use OR, the statements on either side of the connector can be true for the feature to be selected. Suppose you wanted only the values of R-1 and R-2. Would the following equation be correct?

```
CODE = 'R-1' AND CODE = 'R-2'
```

For a feature to be selected, both statements must be true. Because a single feature cannot be both R-1 and R-2, no features will be selected. Now try the same statement using OR, in which either of the statements can be true for a feature to be selected.

```
CODE = 'R-1' OR CODE = 'R-2'
```

A feature with a value of R-1 will be selected, because at least one of the statements is true, and the same with features containing the value of R-2. A feature with a value of I-1 will not be selected because neither of the statements is true.

Your definition query must use OR as a connector and select the following values for the field CODE: R-1, R-1A, R-1L, R-2, TH, and MH.

The definition query you will build will look like this equation:

```
CODE = 'R-1' OR CODE = 'R-1A' OR CODE = 'R-1L' OR CODE = 'R-2'
OR CODE = 'TH' OR CODE = 'MH'
```

3 In the fields box of the Query Builder window, double-click CODE to add it to the expression box at the bottom. Click the equals button (=). Click Get Unique Values and then double-click the value of 'R-1'.

4 To string the next selection onto the query, click the Or button.

5 Continue to build the query. Double-click CODE, click the equals button, and then double-click the value of 'R-1A'.

6 Continue until you have built the entire query, as shown in the graphic. Click the Verify button and let ArcMap validate your query. If there are errors, go back over the query and check each entry. When the query is successfully verified, click OK to close the Verify dialog box. Click OK to close the Query Builder and OK again to close the Layer Properties dialog box.

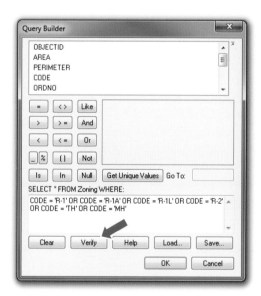

Next, you will take a look at the attribute table of the Zoning Districts layer and see the results of the definition query.

Examine the attribute table again

1 In the table of contents, right-click the Zoning Districts layer and click Open Attribute Table.

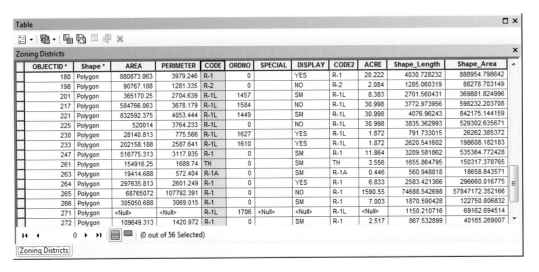

OBJECTID *	Shape *	AREA	PERIMETER	CODE	ORDNO	SPECIAL	DISPLAY	CODE2	ACRE	Shape_Length	Shape_Area
180	Polygon	880873.063	3979.246	R-1	0		YES	R-1	20.222	4030.728232	888954.798642
198	Polygon	90767.188	1281.335	R-2	0		NO	R-2	2.084	1285.060319	88278.703149
201	Polygon	365170.25	2704.639	R-1L	1457		SM	R-1L	8.383	2701.560431	369881.824996
217	Polygon	584766.063	3678.179	R-1L	1584		NO	R-1L	30.998	3772.973956	598232.203708
221	Polygon	832592.375	4053.444	R-1L	1449		SM	R-1L	30.998	4076.96243	842175.144159
225	Polygon	520014	3764.233	R-1L	0		NO	R-1L	30.998	3835.362993	529302.635671
230	Polygon	28140.813	775.566	R-1L	1627		YES	R-1L	1.872	791.733015	26262.385372
233	Polygon	202158.188	2587.641	R-1L	1610		YES	R-1L	1.872	2620.541602	198608.182183
247	Polygon	516775.313	3117.935	R-1	0		SM	R-1	11.864	3209.581862	535364.772428
261	Polygon	154916.25	1689.74	TH	0		SM	TH	3.556	1655.864795	150317.378765
263	Polygon	19414.688	572.404	R-1A	0		SM	R-1A	0.446	560.948818	18658.843571
264	Polygon	297635.813	2601.249	R-1	0		YES	R-1	6.833	2583.421366	296660.016775
265	Polygon	68765072	107792.391	R-1	0		NO	R-1	1590.55	74688.542698	57847172.352166
266	Polygon	305050.688	3069.015	R-1	0		SM	R-1	7.003	1870.590428	122750.806832
271	Polygon	<Null>	<Null>	R-1L	1706	<Null>	<Null>	R-1L	<Null>	1150.210716	69162.694514
272	Polygon	109649.313	1420.972	R-1	0		SM	R-1	2.517	867.532899	40165.269007

Notice now that In the CODE column, all the values you specified are on the list. Remember that the definition query did not delete any data. The values that the definition query did select are only ignored. The process can be reversed simply by removing the definition query.

2 Close the attribute table and save your map document as **Tutorial 1-3.mxd** to the \GIST2\MyExercises folder. If you are not continuing to the exercise, exit ArcMap.

Exercise 1-3

The tutorial demonstrated how to add only certain field values to your symbology editor, and how to build a definition query to restrict the dataset you are working with.

In this exercise, you will repeat the process using land-use codes. Add only the values that are necessary and build a definition query to restrict the dataset to the given list.

- Continue with the map document you created in this tutorial, or open Tutorial 1-3. mxd from the \GIST2\Maps folder.
- Turn off the Zoning Districts layer.
- Add the dataset LandUse.lyr from the \GIST2\Data folder. Turn on the layer.
- In the symbology editor, set the Value field to UseCode and add only the following values. Change the labels to the correct description.
  - A1   Single Family Detached
  - A2   Mobile Homes
  - A3   Condominiums
  - A4   Townhouses
  - A5   Single Family Detached Limited
- Set the colors as necessary and create a layer file of the new symbology.
- Set a definition query to restrict the dataset to only the specified values.
- Add a legend and make any necessary changes to the legend so that it displays all the desired values.
- Save the results as **Exercise 1-3.mxd** to the \GIST2\MyExercises folder.

What to turn in

If you are working in a classroom setting with an instructor, you may be required to submit the maps you created in tutorial 1-3.

Turn in a printed map or screen capture of the following:
Tutorial 1-3.mxd
Exercise 1-3.mxd

Tutorial 1-3 review

The city planner gave you the added task of restricting the dataset to only a specific set of values. You were able to control the display on the map by adding only the values on the list to the symbology editor, and thus to the legend.

A problem with this method is that if you later do any summations or calculations, they will be performed on the entire dataset. One way to solve this problem is to use a definition query. You built a query to select only the features of interest. Always remember that a definition query does not delete data, but merely causes it to be ignored until the definition query is removed.

A query statement must contain, in this order, a field name, an operator, and then a value. You can string these statements together using AND or OR. Using AND means that *both* sides of the equation must be true to select a feature. Using OR means that *either* side of an equation can be true to select a feature.

Study questions

1. What extra properties does the legend have?

2. What query will select the zoning values of I-1 and I-2?

3. What query will select the zoning value of R-1 and acreage greater than 50?

Other real-world examples

It is common to want to work with a subset of your data, without having to make a copy.

A city may have within its boundaries many private water wells, categorized by the aquifer they tap. You may want to restrict your data to a single aquifer to calculate the water removal rate from that water source.

The public works department is responsible for repaving the asphalt and concrete streets in the city. You may want to restrict your data to show only the asphalt streets to give it a total linear mileage of streets paved with this material.

You may have a dataset of all fire department responses in a month and want to do analysis on only the arsons. A definition query on the cause of the fire will accomplish this task.

2
Mapping the most and least

GIS data often has values associated with it, such as the number of employees at a location or the amount of paper produced at a specific mill. These values may come in many forms, such as a total number of something or the ratio of that value compared to the whole. The values can be displayed using various symbols or charts that may themselves indicate the impact of the value. Whatever way it is presented, the data can be mapped to show a relationship.

Tutorial 2-1

Mapping quantities

Mapping the number of things at a given location adds a level of complexity to your map that goes beyond simply mapping where things are. There are several ways to display and interpret quantities in GIS analysis.

Learning objectives

- *Identify discrete features versus continuous phenomena*
- *Work with counts and amounts*
- *Work with ratio normalization*

Preparation

- *Read pages 37–45 in* The Esri Guide to GIS Analysis, *volume 1.*

Introduction

The visual analysis performed in chapter 1 was helpful in presenting data, but it only showed where certain things are. Another aspect of analysis is to show an amount or quantity associated with the features and let the viewer make a comparison.

There are several ways to work with quantities from the data, including counts, amounts, ratios, and ranks. Knowing how many of something are at a specific point or the value of the items or their relationship to other locations can add a whole new dimension to comparing features. For example, knowing where businesses are might show you where the business centers are, but knowing how many workers each business employs can help a transportation planner locate bus stops. The quantity, in this case the number of employees, adds additional information to a simple location map to make it more useful.

When mapping quantities, the data can be mapped in three ways: as discrete features, continuous phenomena, or data summarized by area. Discrete features represent a single item or value. These features can be points, lines, or areas, and the quantity associated with them is for that single feature. You may have points that represent store locations with an associated quantity of annual sales. Lines can represent street segments with the count of cars that cross each segment in a day. Areas might represent lakes with the total volume of water they contain. Each feature is a discrete item and has a value.

Continuous phenomena are usually raster datasets that show values continuously over an area. The weather maps that you see on the local news show rainfall over your whole

2-1
2-2
2-3
2-4

city, even though the rainfall is based on readings from a few rain gauges. You know, however, that rain fell over the entire area, not just at the gauges. Hillshade maps are also continuous phenomena, showing the land slope and elevation over a large area.

The third feature type is data summarized by area. Census data is a familiar type of data that summarizes the number of people who live within an area. Chapter 4 deals with data summarized by area and demonstrates how these datasets are created.

Working with quantities is also a good way to start revealing patterns within mapped data. In chapter 1, you merely showed where zoning categories are located. There were no quantities associated with the data that you could use for meaningful analysis. In this chapter, you will use quantities to start showing patterns. Although mapping the quantities will give a picture from which the viewer can interpret patterns, grouping the quantities into classifications may lead the viewer to a conclusion.

Scenario   An investment opportunity has come your way with a grocery store chain planning to build some specialty stores that target the Hispanic customers in Tarrant County, Texas. These stores have certain brands of food, types of meat, and bakery items that are typically sold in Mexico and are desirable in some US markets. Because Texas was once a part of Mexico, there is still a rich Hispanic heritage in the area, and these markets have become significant supporters of the local Hispanic-themed holidays and festivals.

The chain you are interested in is PePe's Market, which already has five locations in your area. The competition, Buena Vida, has seven locations in the same region. Before you put money into the operation, you want to see if the existing PePe's stores are doing well and are in appropriate markets. You also want to investigate locations for new stores to continue growing the chain.

You will take some existing data about both chains and match it against 2010 Census data to see how the stores fare when compared with the ethnicity of neighborhoods. You will be looking for neighborhoods that will seek out stores with specialty goods imported from Mexico as well as customers that will appreciate your community support.

Data   The first dataset contains the locations of the existing stores. It is a comma-delimited text file with store names, addresses, number of employees, and average monthly sales compiled from the websites of the stores. This information will be turned into a point feature class to use in the analysis.

The second dataset is the 2010 Census block-group-level data. This particular set contains a field for the total population, as well as the number of Hispanics in each block group. A study area is already cut out for you to use, but the census data can be used to repeat the analysis for any area.

The final dataset is the street network data to give the map some reference. It comes from Data and Maps for ArcGIS data.

Create a layer from x,y coordinates

1　In ArcMap, open Tutorial 2-1.mxd.

The layout for your map has been started and a legend added. The map shows Fort Worth, in Tarrant County, Texas, and the surrounding area. The census block-group-level data has been added but is not symbolized.

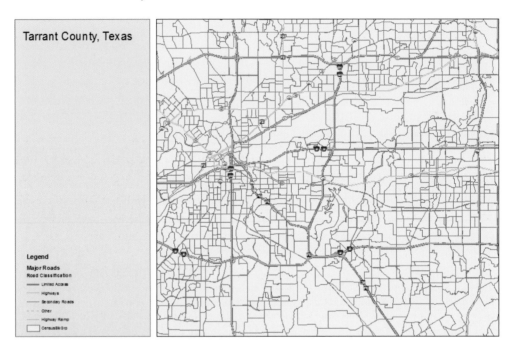

The first thing to do is to add the point data representing the store locations. To add it, you will use a tool named Make XY Event Layer, which can be located using the Search function in ArcMap.

2　On the far-right side of your map document, move the cursor over the Search tab, which opens the Search window automatically. If the Search tab is not present, click the Search Window icon ⬚ on the main toolbar.

3 To find the Make XY Event Layer tool easily, type the tool's name in the Search window and press Enter or click the Search button 🔍.

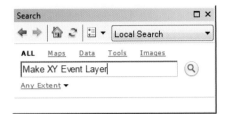

Using the Search window

The Search window is designed as a single tool to find tools, data, maps, and more. The window can be displayed as a temporary pop-up window by pausing your pointer over the Search tab. Moving the cursor away from the Search window will cause it to minimize. Clicking the Search tab produces a more permanent window, which you can close by clicking outside the box. You can dock the window to the toolbars by clicking the pushpin icon 📌 in the upper-right corner of the box. Releasing the push-pin will minimize the window again.

The tutorial uses the ALL setting for the search. The user can also decide to search for specific items by clicking Maps, Data, or Tools. For these searches to perform efficiently, users should add their data folders to the search index through the Options tab.

Once a search is completed, clicking the tool name on the first line opens the tool. Clicking the description on the second line opens a window with a description of the tool. Clicking the third line opens the Catalog window and displays the location of the tool in the System Toolbox. A ToolTip is available for the search results and can be turned on or off under Search Options.

4 Click the Make XY Event Layer tool to run it.

Now you will enter the input data, which is a spreadsheet with the stores' longitude and latitude values.

5 Next to the XY Table text box, click the Browse button, navigate to the exercise data folder, and select FoodStoresHispanic.csv. Click Add.

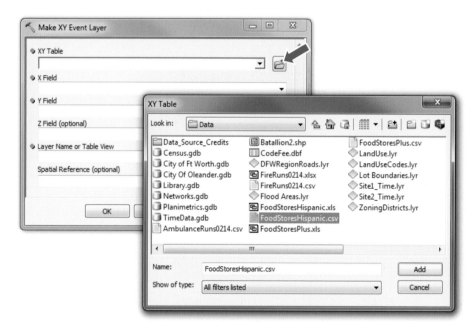

6 Set the X Field and Y Field to the longitude and latitude. Click the down arrow and select bg_long for X Field and bg_lat for Y Field.

Finally, set the spatial reference. Longitude and latitude values are not projected coordinates, but rather unprojected grid values. For ArcMap to project these values on the fly to match the project's spatial reference, identify the spatial reference of the input data. You will use an unprojected world coordinate system named WGS84. If you are new to projections, see "About map projections" in ArcGIS for Desktop Help.

7 Click the Browse button next to Spatial Reference to open the dialog box and then double-click Geographic Coordinate Systems.

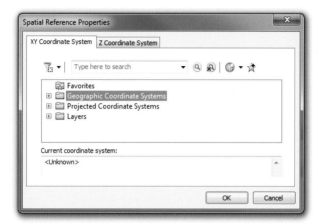

8 Double-click World and select WGS 1984, which is a global projection suitable for coordinates in longitude and latitude. Click OK.

9 When your screen matches the graphic, click OK.

ArcMap will detect the differences between the new layer's coordinate system and the coordinate system of the current map document but reprojects on the fly.

10 Review the Results window and then close it.

This process has made a point layer that shows the locations of the existing Hispanic-themed grocery stores, and has transferred the fields in the table into the new layer as fields. This layer is temporary and exists only in this map document. It will not be saved as a permanent file unless you specifically save it. Next, you will open the attribute table and see what data has transferred with the points.

Review the attribute table

1   Right-click FoodStoresHispanic_Layer and click Open Attribute Table.

The attribute table has the store names and address information, along with the number of employees and the average monthly sales for the year.

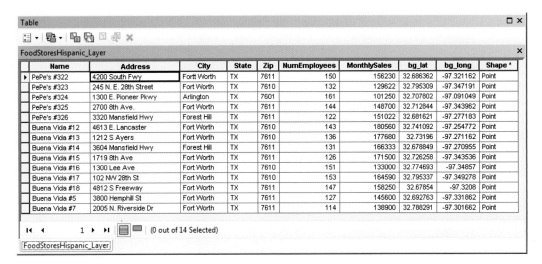

2   Close the attribute table.

For the purposes of investigating whether to invest in this company, you will map average sales to see which stores are doing the best. You could put a label next to each point displaying the average sales figure, but that will make it difficult to do a quick visual comparison between stores. Instead, you will use different-size dots to represent sales. The larger the dot, the higher the sales; the smaller the dot, the lower the sales. A quick scan of the map looking for the largest dots answers the question. A dot comparison is another good example of how informative visual analysis can be.

Set a graduated-symbols classification scheme

1 Open the properties of FoodStoresHispanic_Layer by right-clicking the layer and clicking Properties.

2 Go to the Symbology tab, and in the Show window, click Quantities > Graduated symbols.

3 Click the down arrow next to the Value field box and select MonthlySales.

The symbols will represent the average monthly sales for the past 12 months. The default classification is five classes with a preset size range. Set the sizes slightly larger for the symbols to display better.

4 Change the Symbol Size values from 4 and 18 to 8 and 25.

Try setting the size and then clicking Apply to preview the changes in the map. You may want to experiment with different size and color combinations to get a setup that appeals to you.

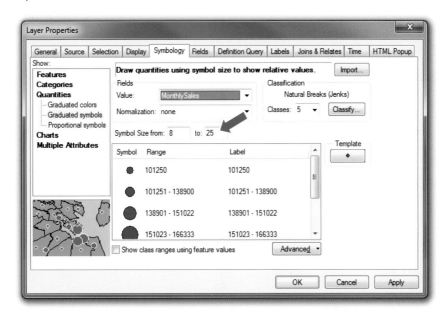

5   When you are satisfied with the results, click OK.

The resulting map shows the store locations symbolized by dots that are sized based on the average monthly sales at each store. Changing the dot size is an example of mapping and symbolizing an amount. With a quick glance at the map, you can see that some stores have higher sales than others. You can also do some additional visual analysis to see how stores relate to the freeway network.

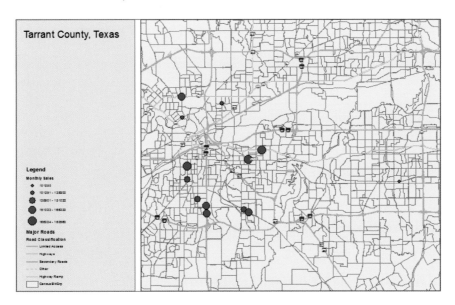

When you sit down and talk with the owners to discuss investing in them, it would be revealing to quiz them on their expansion plans. This map shows the stores to be grouped in certain areas, and you can explore other areas to see if they will be suitable for more stores. You know that the stores cater to the Hispanic population, offering specialty ethnic foods. So if you can find the areas of the county with a high concentration of Hispanics, it might suggest areas for expansion.

Included in this map is the census data for the county. You will look at the attribute table and see if there is a field that can be used to find concentrations of Hispanics.

Examine the census layer attribute table

1   Right-click the CensusBlkGrp layer and click Open Attribute Table.

Population 2010	WHITE	BLACK	AMERI_ES	ASIAN	HAWN_PI	HISPANIC	OTHER	MULT_RACE	MALES	FEMALES	AGE_UNDER5
2273	2061	30	17	2	4	345	129	30	1162	1111	150
1722	1316	78	20	13	4	478	244	47	860	862	161
1183	1081	22	6	6	0	191	39	29	591	592	65
1393	1247	14	4	9	0	212	85	34	694	699	110
1504	1246	104	9	29	1	185	71	44	746	758	76
3778	2807	419	40	86	7	658	298	121	1849	1929	318
6741	4753	1007	37	407	0	1092	322	215	3338	3403	508
846	718	2	11	26	0	121	50	39	441	405	68
2288	2014	94	20	14	1	339	110	35	1126	1162	167
2700	2505	14	29	2	11	255	78	61	1323	1377	175
2841	2603	106	13	21	0	279	65	33	1455	1386	152
2766	2513	47	14	56	0	285	84	52	1396	1370	174
1661	1419	51	17	2	2	341	136	34	854	807	117
814	714	27	8	5	0	122	45	15	405	409	41

(0 out of 5121 Selected)

CensusBlkGrp

2   Move the slider to the right, and you will see a field named Hispanic.

This field represents the number of Hispanics counted in each census tract, for data summarized by area. You can symbolize each polygon according to the count in the Hispanic field.

Summarizing by area is a common way to show data. It is not possible to locate every person and place a point on the map where each one lives—the data would not only be complex, but personally revealing about individuals. It is more convenient, and a better safeguard of privacy, to set small study areas and summarize the data for that area. For census data, you can see how many people live on a certain block, but not gain personal information about an individual household.

3   Close the attribute table.

Set the classification scheme

1   Right-click the CensusBlkGrp layer and click Properties. Go to the Symbology tab, and in the Show window click Quantities > Graduated colors.

Note: the graphics show the default color ramp; your colors may be different because ArcMap remembers the last color ramp chosen.

2   In the Value field drop-down list, select Hispanic.

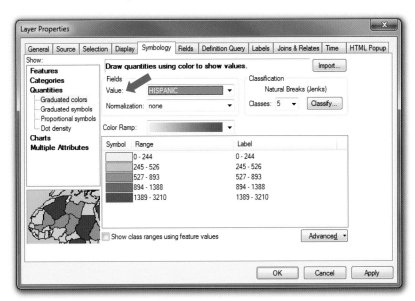

You will be using the count of Hispanics in each census tract to assign a color. The tracts with the darkest color will have the largest Hispanic population counts, and the lighter colors will represent the lower counts.

3 Click Apply to preview the results. If necessary, move the dialog box so you can see the map. You can try different color ramps, and when you are satisfied click OK.

You accepted the default settings for the classification. The default classification method is called Natural Breaks and displays five categories. This chapter will show more about this method later, but for now it is sufficient to recognize that the darker color represents a larger count, and the lighter color represents a lower count.

Using visual analysis, you can see that the stores seem to be grouped near the polygons with the darker color. Stop for a moment and think about what you are seeing. The darker areas have a large number of Hispanics, but does this number represent a large percentage of the population in that census tract? A tract may score high on this map with a count of 2,500 Hispanics, but what if there are 15,000 people in the census tract? That count is only 16.7 percent. The start of this tutorial said to look at concentrations, or areas where there are more Hispanics as a percentage of the total population. You can calculate the percentage by dividing the Hispanic count in each tract by the total population of each tract. This process is called *normalization*. You are looking for the population of Hispanics normalized by the total population. For these purposes, you can substitute the words "divide by" for "normalize by." So to normalize by total population means to divide by total population. This example of mapping and symbolizing a ratio can be achieved by modifying the classification scheme.

2-1
2-2
2-3
2-4

Set a normalization field

1 Open the properties of the CensusBlkGrp layer and go to the Symbology tab. Click the down arrow for Normalization and select Population 2010.

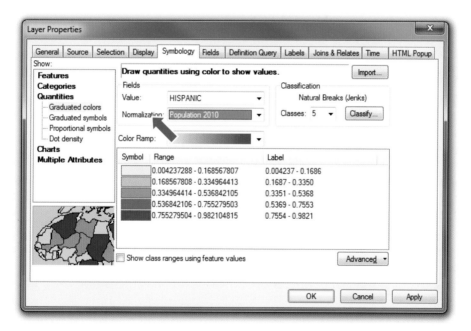

2 Click OK to accept the changes.

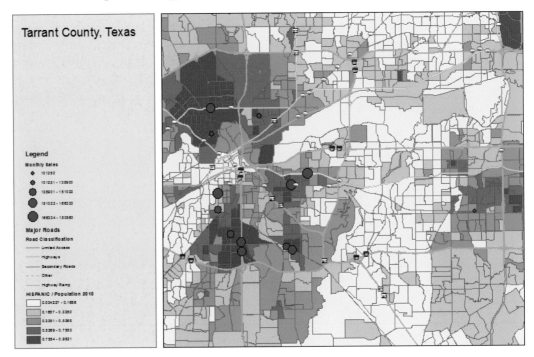

Wow, what a difference! Now you are looking at areas that are predominantly Hispanic and would be likely to shop at your specialty grocery store and benefit from the community support the store provides. There are also some areas where the stores are doing well, and there seems to be room for more stores. The map also gives insight into why the easternmost store is not doing so well. Although the count of Hispanics seemed high, it is not in an area of high concentration. So a large area with a large number of people is not as desirable as a small area with a large number of people. The ratio of potential customers is lower. It is important to be aware of what you are asking for in your analysis. This scenario demonstrates that showing count versus concentration can produce two very different results.

3 Save your map document as **Tutorial 2-1.mxd** to the \GIST2\MyExercises folder. If you are not continuing to the exercise, exit ArcMap.

Exercise 2-1

The tutorial showed how to create an XY event layer from a comma-delimited text file that contains longitude and latitude values. It went on to demonstrate how to symbolize a dataset with graduated symbols and colors.

In this exercise, you will repeat the process using different datasets. You are also looking at investing in a different type of specialty grocery store. This one caters to high-end clients that are looking for more exotic foods from throughout the world. Each store has a product selection rank, representing how many exotic food selections it carries. You will also display store locations over income-level data taken from the US Census.

- Continue using the map document you created in this tutorial, or open Tutorial 2-1.mxd from the \GIST2\Maps folder.
- Turn off the CensusBlkGrp layer. If you are continuing with a completed map document from Tutorial 2-1, you will also need to turn off the FoodStoresHispanic layer.
- Make an XY event layer from the file FoodStoresPlus.csv in the data folder.

The data is also in longitude-latitude coordinates and needs a spatial reference that will accommodate them.

- Symbolize the FoodStoresPlus point layer with graduated symbols using the ProductSelection field.
- Adjust the scale so that all the new points in the FoodStoresPlus layer are visible.
- Add the CensusBlkGrpIncome layer to your map document (found in \Census.gdb \DFWRegion).
- Symbolize the CensusBlkGrpIncome layer with graduated colors using the median Household Income field (determined by the Census 2010 metadata) to represent median household income.
- Change the titles, colors, legend, and so on, to make a visually pleasing map.
- Save the results as **Exercise 2-1.mxd** to the \GIST2\MyExercises folder.

What to turn in

If you are working in a classroom setting with an instructor, you may be required to submit the maps you created in tutorial 2-1.

Turn in a printed map or screen capture of the following:

Tutorial 2-1.mxd
Exercise 2-1.mxd

Tutorial 2-1 review

Showing quantities on your maps provides another level of complexity beyond simply showing where things are. You can visualize counts, amounts, and ratios with other locational data shown as an overlay.

The graduated-symbols classification scheme is a great way to symbolize quantities associated with point data.

Remember to let the dot sizes tell the story, and do not confuse things by trying to make the dots different colors.

If the map viewers realize all the purple dots represent different values of the same item, they will get a much better understanding of the data. A map full of different-colored dots that vary in size is confusing and will not read well to viewers.

It is also important to remember what the quantities represent and how to interpret them. As you saw with the normalization feature in the layer properties, you can create ratios on the fly. You used the total number of Hispanics and divided (or normalized) by the total number of people of all ethnic groups. Always make sure when setting a normalization field that it matches a real mathematical equation you can perform. The equation mimicked here was to get the percentage of the total (part/total × 100, or Hispanic population divided by total population × 100).

Another normalization selection built into the layer properties is called Percent of Total. This tool divides the number of Hispanics in a particular tract by the total Hispanic population, or the sum of the value field. Try changing the percentage on your map and notice the difference in the results. It is fine to use this method, but make sure that you understand what the map is showing (the percentage of the total Hispanic population that resides in each tract) before you use it.

There is more discussion on classification methods later in this book, but it is interesting to note that these maps read well with the default settings, and can be labeled with such basic text as low/medium/high. The actual numbers are not as important as the relationship between the numbers.

Study questions

1. Read in ArcGIS for Desktop Help the difference between graduated symbols and proportional symbols. When do you use one over the other?

2. Will changing the number of classes or the display colors cause the map to show a different answer?

3. What are the responsibilities of the person creating a map to display the data without manipulation?

2-1
2-2
2-3
2-4

Other real-world examples

A police department commonly maps locations of auto accidents and uses graduated symbols to display multiple accidents at a single location. This symbolization highlights problem intersections.

US Census data is a treasure chest of quantity data. For a complete understanding of all the utility that census data can provide, see *Unlocking the Census with GIS* by Alan Peters and Heather MacDonald (Esri Press, 2004).

Traffic counts are routinely done for stretches of roads, and then linked with the GIS data. The streets are mapped with graduated symbols: the larger the symbol, the higher the count; the thinner the street, the smaller the count.

Tutorial 2-2

Choosing classes

You have many different classification methods to choose from, and the right choice can make or break your map. The data distribution diagram, sometimes known as the histogram or frequency distribution chart, will help you make the right choice.

Learning objectives

- *Understand classification schemes*
- *Understand data statistics*
- *Understand data distribution characteristics*

Preparation

- *Read pages 46-55 in* The Esri Guide to GIS Analysis, *volume 1.*

Introduction

When mapping quantities, you can divide the data into classes, or ranges of values, to show on the map. The methods for putting data into classes, or classification, include natural breaks, quantile, equal interval, and standard deviation. A combination of the desired look of the map and the distribution of the data is used to determine which method to use.

The most common, and the default choice in ArcMap, is the natural breaks method, using the Jenks algorithm. Mathematician George Jenks developed the method of finding natural groupings of data, and setting classifications based on those groupings. This method groups similar values and maximizes the differences between classes. For most studies, the Jenks natural breaks method is the preferred choice.

The quantile classification method deals directly with the number of features in each class. The number of features is divided by the number of classes you specify, and the resulting quantity is placed in each class. The quantity of each class is the same. This classification works well if you want to show just a certain percentage of the results—for instance, the top 20 percent of the values. It also works well if the data is evenly distributed across the entire value range. It is not desirable for data in which the values tend to group together or in which the measured area of the features is vastly different. Values that are not similar may be placed in the same grouping.

The equal-interval classification involves the size of the classes. The total range of the data values is divided by the number of classes you specify. Values are then placed in the classes with no regard as to how many values fall within each range. This method works well if the range of values is familiar to the audience, such as percentages. But in some instances, it may place all the values in one or two classes, leaving other classes empty.

The standard-deviation classification shows how much a value varies from the mean. The standard deviation is calculated for each value, and the result is classified as above or below the standard deviation. The data must be tightly grouped in a classic bell curve for this classification to be effective. Otherwise, you get a significant number of features four and five standard deviations from the norm, which will not show any groupings. The drawback to this classification is that the actual values are not shown, only their relationship to the mean. Very high or very low values can affect the calculation and skew the results.

Step one in deciding which classification to use is to view the data distribution diagram for the dataset. Evenly distributed data might be a candidate for a quantile or equal-interval classification. Data that is grouped tightly or for which you want to show the median values might be best shown using a standard-deviation classification. Data that has one or more distinct groupings is best shown using the Jenks natural breaks classification.

After choosing the classification method, determine the number of classes. The larger the number of classes, the less change between the class values, thus making changes subtler and harder to detect. If you need a more complex map for a sophisticated audience, using more classes may work well.

Conversely, fewer classes will greatly simplify the data display. A quick-glance analysis for a general audience may call for a simpler map. Sometimes a simple "low, medium, high" split is sufficient.

As you gain more experience with different datasets, classification methods, and viewing audiences, you will begin to develop a feel for which combination works best.

Scenario   The city planner has given you several datasets and has asked you to determine the best classification method for each one. He is not sure what he wants, so you must show him what can be done, listing the classification method and why it will or will not work for this analysis.

Data   The first dataset is the census block-group-level data for Tarrant County, Texas. The fields Median Age and Population 2010 contain the counts for each census tract.

The second dataset is the parcel data for the City of Oleander, Texas. You will display the values in the YearBuilt field, setting up a classification to show the age of housing.

Examine the attribute table of the census layer

1 In ArcMap, open Tutorial 2-2.mxd.

You get the familiar map of Tarrant County that contains the census data. You will look at the data fields one by one and try to determine the best classifications to use.

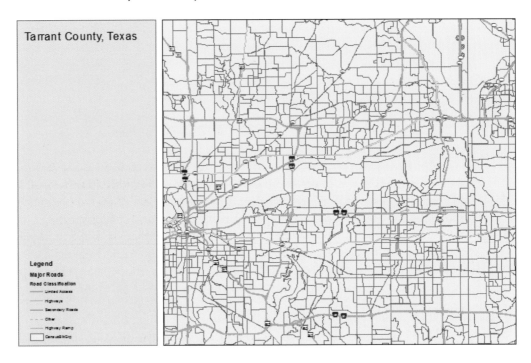

2 Right-click the CensusBlkGrp layer and open the attribute table.

3 Right-click the Population 2010 field and click Statistics.

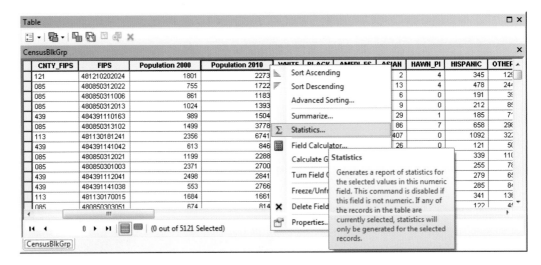

The statistics box opens, showing various statistics about the data contained in this field. The interesting part is the Frequency Distribution diagram on the right.

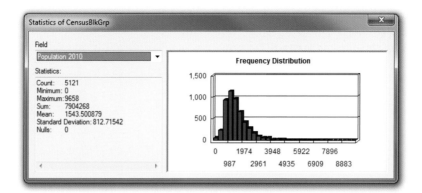

The numbers across the bottom represent the range of values in this field, and the numbers along the side are the number of features that have the same value. What you want is to visualize a curve over the chart display to see how well the data is distributed across the range of values. You see that the data is a regular curve, weighted to the lower end of the scale. As the first test, you will classify the data using the Jenks natural breaks method and let it find the natural groupings of data.

4   Close the Statistics window and the attribute table.

Test the classification scheme for total population

For a support map in a series of analyses showing population characteristics, you will display the population total so that is easy to interpret. The map should give the viewer an idea of where the largest population centers are, where the smallest population centers are, and provide a range of values in between. If you show only areas of high or low population, there will not be a good frame of comparison between the values.

2-1
2-2
2-3
2-4

1 Right-click the CensusBlkGrp layer and open the properties. Go to the Symbology tab, and in the Show box click Quantities > Graduated colors. Set the Value field to Population 2010.

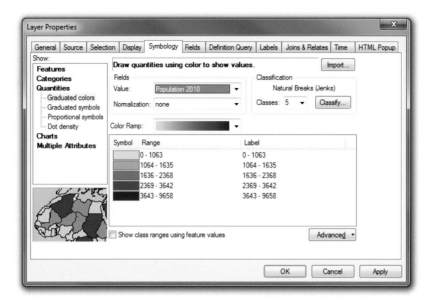

The default classification of Jenks with five classes is selected. In many instances, you can use the Classes drop-down list to adjust the number of classes and continue using the natural breaks classification. But the city planner wants an explanation of the choice, so look into it further.

2 Click the Classify button to open the Classification dialog box.

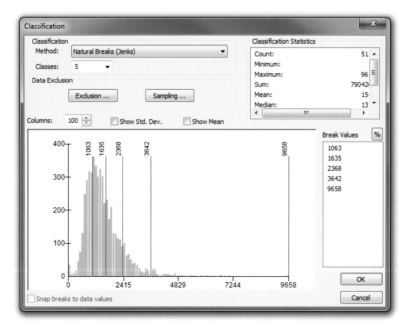

The data values are grouped to the left of the chart. You can see that the gaps between the classification breaklines in blue are smallest at the left and get larger as the values increase. When you view the breaks on the map, you expect to see many polygons representing the first two classes, and only a few polygons representing the rest of the data.

3 Click OK. Change the color ramp if you want, accept the default classification settings, and click OK to close the Properties dialog box. Examine the results on the map.

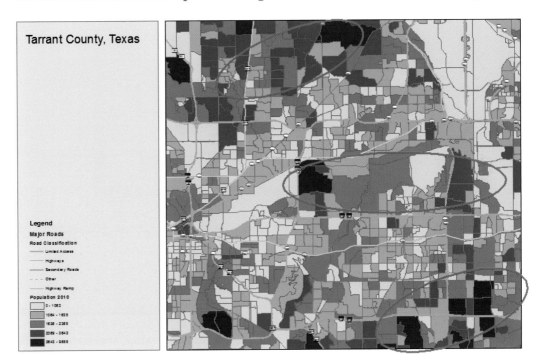

It is easy to see the groupings of data that correspond to the classifications you set. Some circles are drawn in the graphic for emphasis, but they will not appear on your map. All ranges are represented on the map, and the data is easy to interpret. So now, you can tell the city planner that the Jenks natural breaks classification works well for population data.

4 Print or save a screen capture of this map to the MyExercises folder.

Next, you will look at the Median Age field and choose a classification for it.

Test the classification scheme for median age

As with the total population map, the map of median ages should highlight both the high and low areas, as well as the medium-range values for comparison.

1 Right-click the CensusBlkGrp layer and click Open Attribute Table. Right-click the Median Age field and click Statistics to open the Statistics window.

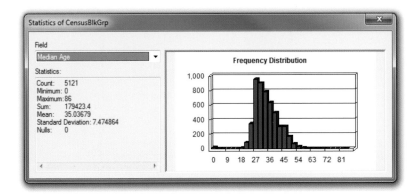

Draw an imaginary curve along the top of the chart, and you will get the classic bell curve with the data centrally weighted. You will try a quantile and an equal-interval classification on this field.

2 Close the Statistics window and the attribute table.

3 Right-click the CensusBlkGrp layer and open the properties. Change the Value field to Median Age.

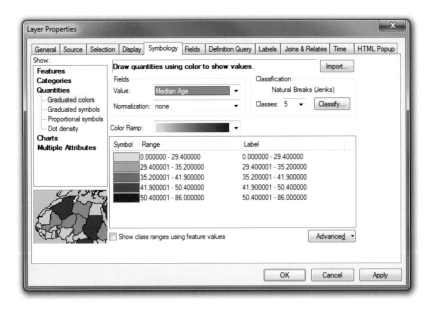

4   Click the Classify button and change the classification method to Equal Interval.

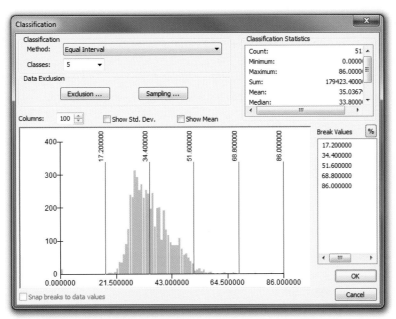

As the name implies, the intervals between the classification breaklines are equal. You can verify this equality by looking in the Break Values display. The values are approximately 15 units apart. Next, look at how many features are in each group.

5   Click the first value in the Break Values box.

At the bottom of the dialog box, you can see the number of features in the first class.

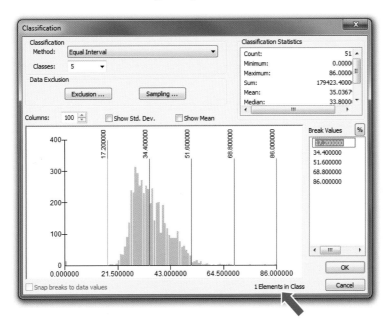

6 One by one, click the break values and look at the resulting number of features in each class. Click OK and then OK again to close the dialog boxes and see the results of this classification on the map.

You will find that the second and third classes contain most of the features. When displayed on the map, only a few of the lowest classifications are shown. Most of the map is represented by only two of the colors. The data does not lie; these values are true. But this classification does not show any distinct groupings. This classification is not suitable, so you will try another one.

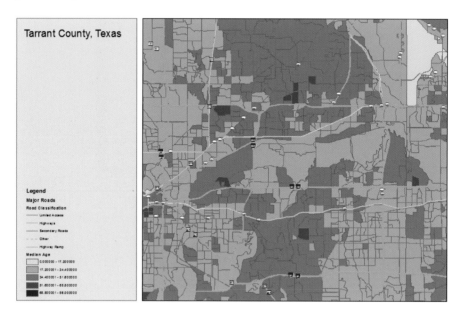

7 Open the layer properties of CensusBlkGrp again and click the Classify button. Change the classification method to Quantile.

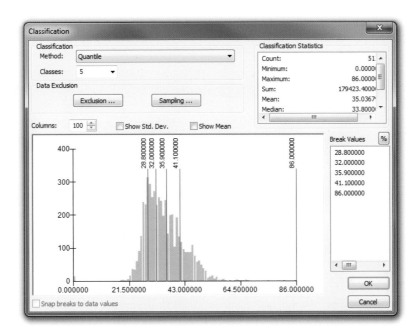

By looking at the Break Values field, you can see that the first class range is large. The next few class ranges are small, and then the last class range is large again.

8 Click each of the break values and note how many features are in each class.

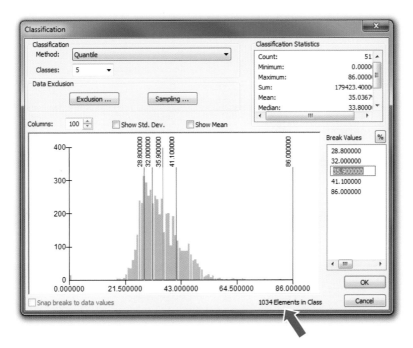

Quantile means that each class has close to the same number of features, which can be seen by examining the break value numbers. There are around 1,030 features in each group.

9   Click OK and then OK again to close the Properties dialog box and view how the classification setting changes the map display.

Some good groupings are apparent with this classification, and all the value ranges are represented. The quantile method is appropriate for this dataset.

10   Print or save a screen capture of this map to the MyExercises folder.

Now, you will look at the age of houses in Oleander. Each parcel has a year of construction for the house, if there is one on the property. By looking at the data, you might be able to see groupings of years in which there might have been a construction surge. The question is, which classification scheme will be the best to display these groupings?

Analyze year-built data

The data that shows the year in which each house was constructed can help pinpoint areas in need of renewal. The data must be shown so that it can be easily understood, and matched to conventions for date-based data.

1 On the main menu, click Bookmarks and select the City of Oleander bookmark. Change the map title to **City of Oleander**.

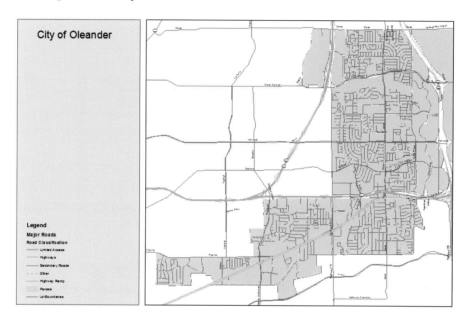

The data is a parcel map of Oleander. Right now, the parcels are all the same color, but you will change the classification scheme to show the year of construction for each house. Before you decide on which classifications to consider, you will look at the Frequency Distribution diagram.

2 Right-click the Parcels layer and open the attribute table. Right-click the field YearBuilt and click Statistics.

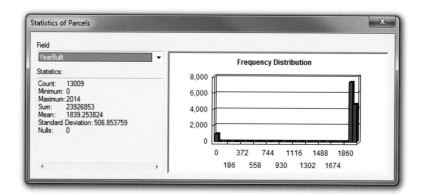

The distribution diagram looks odd. Some values are at zero; the rest are in the upper ranges. What you are seeing are the values for the parcels for which there is either no structure, or the year of construction is unknown. These values, called *outliers*, are skewing the data to the low end of the scale.

3 Close the Statistics window and the attribute table.

If you did a definition query to remove all the outliers, those parcels will not be included in the Frequency Distribution diagram and the data will look better. But those parcels also will not display on the map, leaving many gaps. Instead, a feature in the properties instructs ArcMap to ignore outliers in the classification scheme.

Note: this dataset is large, and ArcMap will likely reach the maximum number of samples used to set the symbology groupings. The default sample of 500 features will work fine for this dataset, and you can ignore the warnings. If a larger sample size is needed, the value can be changed by clicking the Sampling button on the Classification dialog box.

4 Right-click the Parcels layer and open the properties. Set the Show window to Graduated Colors, change the Value field to YearBuilt, and click the Classify button.

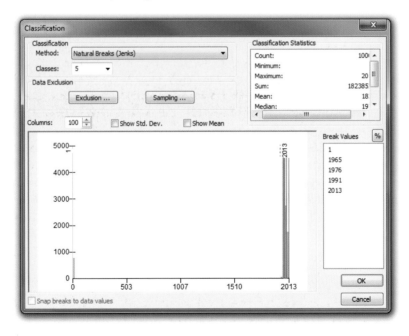

You will use the Data Exclusion function to remove all the values that are less than 1,900. Then you will review the Frequency Distribution diagram again and decide which classification method to use.

5 Click the Exclusion button. Double-click the field YearBuilt, click the < symbol, and then type a space and the number **1900**. When your dialog box matches the graphic, click OK.

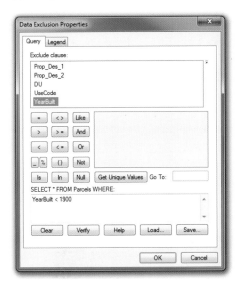

Now look at the difference in the Frequency Distribution diagram. The features that meet the Data Exclusion query are no longer considered in the Frequency Distribution diagram or used to determine the classes and values for the classification.

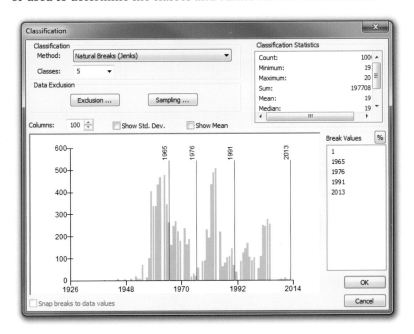

The data spans from 1926 to 2014, or 88 years. The natural breaks method does not classify the data in a manner that works well with dates. The ranges span anywhere from six to 15 years, which makes comparisons difficult. What if you set the intervals to 10 years to make them easier to understand? To do that, set the classification method to Equal Interval and the number of classes to nine. Dividing the 88-year span into nine equal intervals makes the intervals 10 years each.

6 Change the classification method to Equal Interval, and set the number of classes to 9. Examine the Frequency Distribution diagram. Click OK.

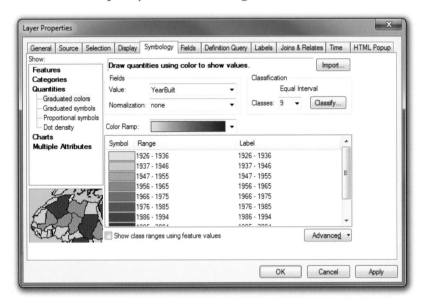

Change the symbol outline

1   Right-click one of the colored boxes in the Symbol column and click Properties for All Symbols.

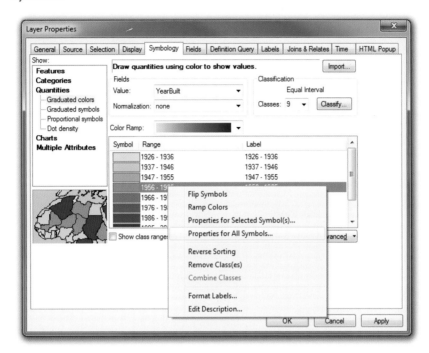

2   In the Symbol Selector, set Outline Width to 0. Click OK and then OK again to close the properties.

The result clearly shows the progression of construction across Oleander over the past 88 years in time ranges that make sense to a broad audience.

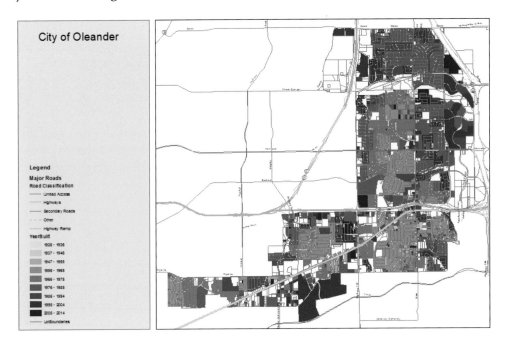

But the map can be improved. Although it makes sense for the class intervals to be 10 years, it makes more sense for the intervals to start and stop on the decade. The automated classification methods will not differentiate by decade, but you can do it manually.

Set a manual classification

1 Right-click the Parcels layer and open the properties. Click the Classify button. Change the number of classes to **10**.

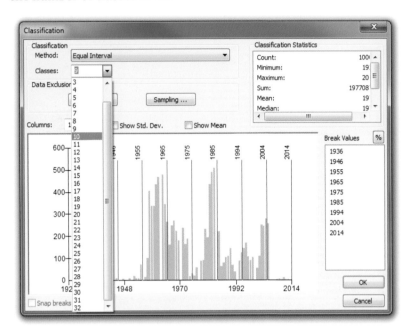

2 In the Break Values box, type new values representing the decades, starting with **1929** and ending with **2019**.

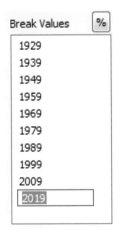

3 Click OK and then OK again to close the dialog box.

Now the ranges in the legend represent whole decades, fitting a standard convention of how people think of years. Everyone who looks at the map will easily understand this time range.

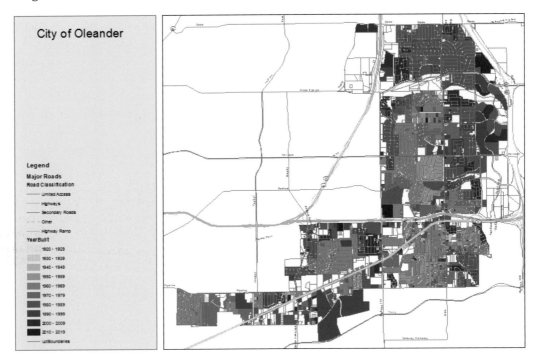

4 Print or make a screen capture of this map. Save your map document as **Tutorial 2-2.mxd** to the \GIST2\MyExercises folder. If you are not continuing to the exercise, exit ArcMap.

Exercise 2-2

The tutorial showed various classification methods and why they were chosen for particular datasets.

In this exercise, examine the field DU in the Parcels layer. This field represents dwelling units. Exclude the zero values, which are vacant lots. Then select a classification method that will give the viewer an idea of areas with one or two dwelling units per parcel, areas with a midrange of dwelling units per parcel, and areas with 50 or more dwelling units per acre.

- Continue with the map document you created in this tutorial, or open Tutorial 2-2.mxd from the \GIST2\Maps folder.
- Go to the City of Oleander bookmark.
- Examine the Frequency Distribution diagram for the DU field in the Parcels layer.
- If necessary, change the symbology scheme to graduated colors.
- Change the Value field, classification method, and number of classes as necessary.
- Add a text box to the map document and describe what affected your choice of classification method.
- Change elements such as the titles, colors, and legend to make a visually pleasing map.
- Save the results as **Exercise 2-2.mxd** to the \GIST2\MyExercises folder.

What to turn in

If you are working in a classroom setting with an instructor, you may be required to submit the maps you created in tutorial 2-2.

Turn in a printed map or screen capture of the following:

> **The three maps you created in tutorial 2-2**
> **Exercise 2-2.mxd**

2-1
2-2
2-3
2-4

Tutorial 2-2 review

Although you should strive for realism in your maps, there is no practical way to display all the quantities that a field might contain. They must be simplified, or classified, to show a model of reality. Many ways exist to classify data into groups, and you saw that not all classifications represent the data correctly. Some methods put all the features into one class, leaving a lot of blank space on the map. Others spread the data so evenly that no true pattern emerges.

The number of classes can also make a difference. Too many classes might confuse the viewer, and too few might make the map too simple to derive any meaningful answers. You also saw how to set the classes manually to conform to a more easily understood convention of showing years by decade, even though the automated classifications produce a nice result.

Choosing a classification all starts with examining the distribution of the data. The Frequency Distribution diagram provides a chart of the values and points the way to a classification method that is more likely to produce the desired result.

Study questions

1. What is the difference between an equal-interval and a quantile classification?

2. What does the Frequency Distribution diagram show?

3. Describe examples of outliers. How can you deal with them?

Other real-world examples

A quantile classification might be used to display a dataset as percentages of the whole. Making four classes with the same number of features in each class will result in breaks of 25 percent.

An equal-interval classification might be used to show individual age, grouping the values by 10 to represent 10 years.

A natural breaks classification can be used to classify a dataset of many lower-value houses, a few medium-value houses, and many high-value houses. Because the dataset has two natural groupings, the classes should be set to highlight them.

Tutorial 2-3

Creating a map series

A single map showing population breakdown was sufficient in tutorial 2-2, but sometimes you must look at data from multiple attribute fields. An effective way to look at data is using a map series. Each map displays different attributes in a similar way, allowing the viewer to do analysis across the maps.

Learning objectives

- *Create multiple map displays*
- *Look for patterns*
- *Work with multiple attributes*

Preparation

- *Read pages 56-60 in* The Esri Guide to GIS Analysis, *volume 1.*

Introduction

The 2010 Census Data includes a series of fields that represent ethnic and racial breakdowns. Using graduated colors, you can display only one of these features at a time. So it is easy to see concentrations of a particular ethnic or racial group, but it is difficult to see how that concentration compares with other groups.

One way to solve this problem is to create a map series. You can make a new map for each study group, symbolizing the quantities in a like manner. The viewer then examines all the maps side by side and compares the data. One important thing to remember when doing a map series is that the data must be classified the same on all the maps for the viewer to easily make comparisons. If you showed Asians on one map with three classes and Pacific Islanders on a second map with five classes, the viewer will not be able to compare the highest values of the two maps. The same color ramp might be used to represent the two datasets, but the colors will represent different ranges of concentrations.

Scenario　　You are still interested in the investment opportunity presented by the grocery store chain, but you want to see if the idea would transfer to other study groups. You will create a map series showing different ethnic groups across the county and compare them. You are looking for another ethnic or racial group that might lend itself to the development of a new chain of specialty grocery stores.

Data　　You can continue to use the 2010 Census block-group-level data. In the attribute table, there is a series of fields that represent different groupings of counts that you can map.

Create the first map in the series

1 In ArcMap, open Tutorial 2-3.mxd.

Once again you get the familiar map of Tarrant County that contains the unsymbolized census data.

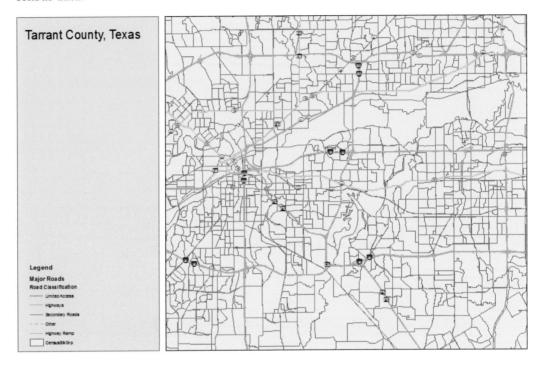

The goal is to create a series of maps that display various ethnic groups. You want to make the maps have the same "look and feel" so that the viewer can easily compare the data across the series. You also want to make sure that the maps cover the same area at the same scale and, most important, have the same ranges on the legend. For one study group, you may get a maximum count of 2,000, while for another study group you may get a maximum count of 20,000. If both of these values are displayed with the same intensity of color for the top range of values, the viewer might make the false assumption that the total numbers of each group are similar.

For this series, you want to create maps that show the total population, the Hispanic population, and the black population. These three maps can be viewed next to each other to give a good idea of how these groups are dispersed across the county.

Make the map of the total population first, and then use the classification ranges for that dataset to display the other datasets. This display will give you the best cross-map comparison.

2 Open the properties of CensusBlkGrp and go to the Symbology tab.

3 Click Quantities > Graduated colors and set the Value field to Population 2010.

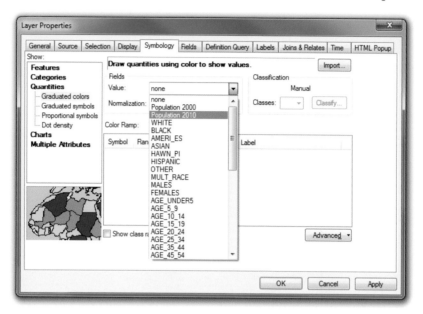

4 Accept the default classification, change the color ramp if you want, and click OK.

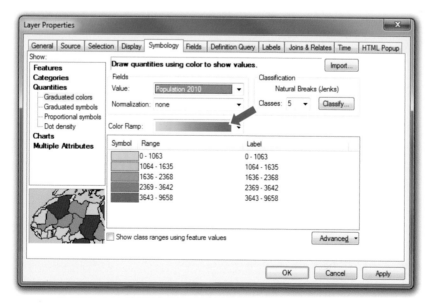

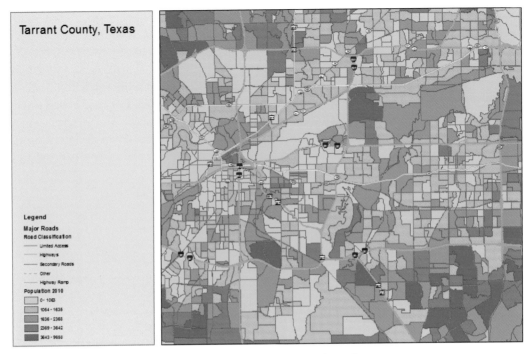

Next, you will change the title of the map to reflect the data shown.

5 Right-click the title, click Properties, and change the title to **2010 Total Population.** Click OK.

6 On the main menu, click Insert > Title.

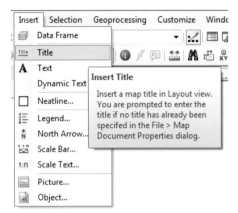

ArcMap inserts a title text block at the top of your layout. By default, ArcMap inserts code into the text box to display the map document name.

7 In the Properties dialog box, delete the code in the text box and replace it with **Tarrant County, Texas**. Drag the text to the left of the map and place it under the map title.

2010 Total Population
Tarrant County, Texas

Do not worry if the text looks too large, or if it is not centered under the other text. There are some tools you can use to fix that.

8 If necessary, add the Draw menu to your map document, and use it to set the text size for the map title to 20.

If you are not sure which button will change the Font Size box, point to the different icons on the menu with your pointer and read the ToolTip. Next, you will fix the alignment.

9 Press and hold Shift and select the map title. You have now selected both text boxes.

10 Right-click the text and click Align > Align Center. Then move the text back to the top of the gray box.

Another way to accomplish the alignment is to use text formatting in the Text editing box. Formatting also allows you to do a multiple-row title with different fonts or text sizes. Read more by clicking the About formatting text button in the Text properties box.

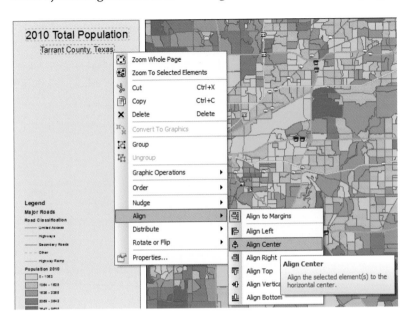

11 Print the map or go to File > Export Map to export it as an image or PDF file and save to the \GIST2\MyExercises folder.

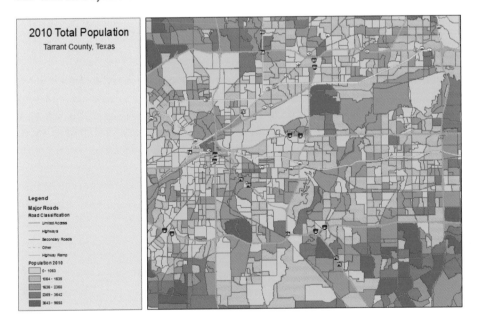

Next, you will duplicate the process to make the other two maps in the map series. First, you will copy the census data file so that you can symbolize it for the next field.

Create the second map in the series

1   Right-click the layer CensusBlkGrp and click Copy.

2   Right-click the data frame Tarrant County and click Paste Layer(s).

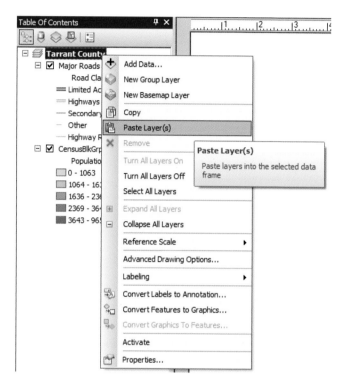

The layer is pasted to the table of contents, but it looks exactly like the other layer and has the same name. Change the name of the layer to avoid confusion, and the symbolization to reflect a different set of data. Remember that it is important to use the same classification scale as the total population, so you will use the Import button in the Symbology Editor to match the other layer.

2-1
2-2
2-3
2-4

3 Open the layer properties of the copied dataset and go to the General tab. Type **Hispanic Population** as the name of the new layer.

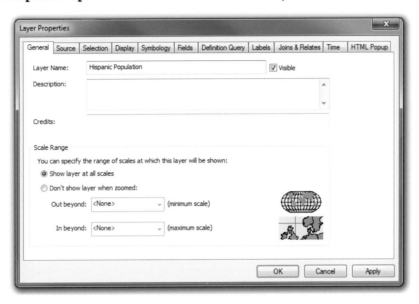

4 Go to the Symbology tab and click the Import button.

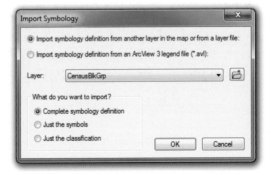

5 There is only one choice of layers to import from, so click OK.

6 Change Value Field to HISPANIC to use this other field for your symbolization and click OK.

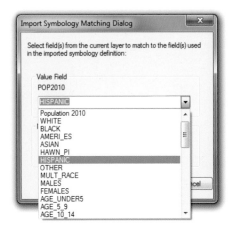

Notice that the value field is now HISPANIC, and the ranges for the classification are the same as the ranges for the total population. If you wanted to look at just one ethnic group, it is fine to allow ArcMap to automatically set these ranges. But to compare different groups, they must reference the same ranges.

7 When you are finished reviewing the settings, click OK.

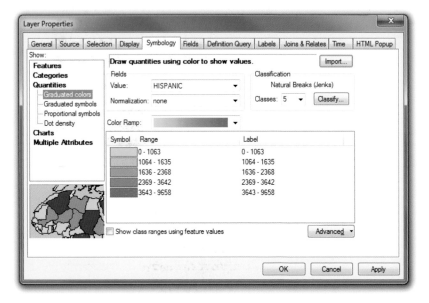

8 Turn off the CensusBlkGrp layer.

Now you will change the map title.

9 Double-click the map title to open the Properties dialog box and type **2010 Hispanic Population**. Click OK to accept the changes.

10 Print the map or export it to an image or PDF file and save to the \GIST2\MyExercises folder. Look at both of these maps side by side. On the Hispanic population map, you do not see the darkest color showing a large population, but you still see some groupings. And you are using the same reference to compare the values.

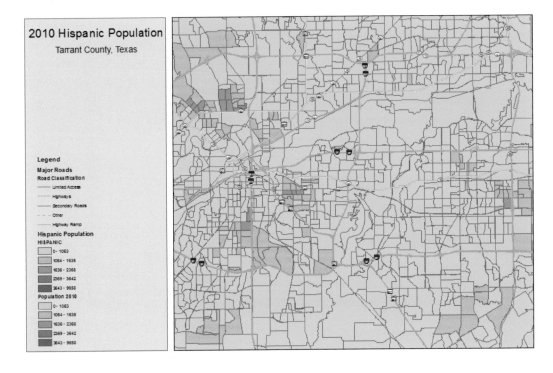

YOUR TURN

Repeat the process and create a map using the field **BLACK** to show the population of blacks. Make sure to import the symbology and retain the same classification ranges. Change the title to **2010 Black Population**. When you are finished, print or export the map to an image or PDF file and save to the \GIST2\MyExercises folder.

2-1
2-2
2-3
2-4

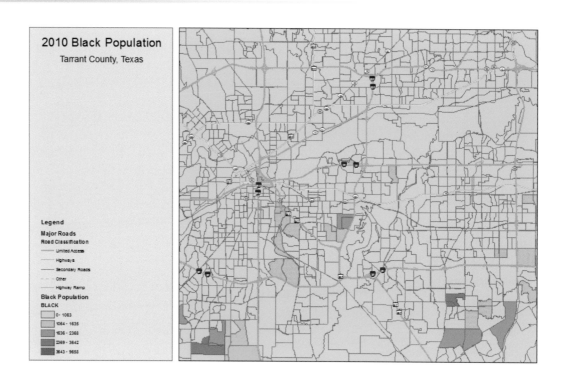

11 Save your map document as **Tutorial 2-3.mxd** to the \GIST2\MyExercises folder. If you are not continuing to the exercise, exit ArcMap.

Exercise 2-3

The tutorial showed the process of making a map series to show the ethnic breakdown in Tarrant County, Texas. The key is to use the same classification values on all the maps so that you can make comparisons between them. The maps show the count for various ethnic groups.

In this exercise, you will repeat the process to show a concentration of groups rather than the count.

- Continue using the map document you created in this tutorial, or open Tutorial 2-3.mxd from the \GIST2\Maps folder.
- Symbolize CensusBlkGrp with graduated colors based on the HISPANIC value field. Make sure to use the same ranges as before.
- Normalize the values by the total population using the Population 2010 field.
- Insert a text box and describe the difference between using a count and using a concentration.
- Change elements such as the titles, colors, and legend to make a visually pleasing map.
- Save the results as **Exercise 2-3.mxd** to the \GIST2\MyExercises folder.

What to turn in

If you are working in a classroom setting with an instructor, you may be required to submit the maps you created in tutorial 2-3.

Turn in a printed map or screen capture image of the following:

The three maps you created in tutorial 2-3

Exercise 2-3.mxd

Tutorial 2-3 review

Since you wanted to compare the values across the different maps, it was important to be careful to use the same classification ranges. If you are only interested in each map individually, you can reset the classification ranges for each map. But to make cross-map comparisons, use the same base classification ranges in each map.

This map series shows the differences between various data values captured at the same point in time. Another type of map series shows how a single data value changes over time. This topic is covered in chapter 6.

2-1
2-2
2-3
2-4

Study questions

1. Why is it important to keep the classification ranges the same in a map series?

2. Do the same rules apply for classifying points, lines, and polygons?

3. Can you show several of the field values together on the same map?

Other real-world examples

The Texas Department of Parks and Wildlife may want to map bird counts at each state park in a given month. To compare which parks have the highest concentrations of birds, it will make a map series showing each bird type normalized by the total bird count.

The Texas Department of Transportation may have data that shows both auto and tractor traffic counts. Each of the traffic count datasets may be used to create a concentration map. The two maps produced will have the same classification range values so that the numbers can be compared on the same scale.

Tutorial 2-4

Working with charts

You saw that a map series is helpful in comparing similar values across several maps. But the solid color fill prevented you from showing more than one field value on a map. Charts, however, will allow you to show several fields at the same time.

Learning objectives

- *Create charts*
- *Select fields for charts*
- *Compare field values*

Preparation

- *Read pages 61-62 in* The Esri Guide to GIS Analysis, *volume 1.*

Introduction

A chart can display several attributes at the same time and allow the user to quickly compare values against the same scale. The charts can easily be constructed from the attribute tables of your data and added to your map layouts. However, as you will see in this tutorial, charts have some visual limitations.

Creating charts on a large-area map with many small polygon areas may make the charts difficult to see, and the software will automatically remove some of the overlapping charts for clarity. Charts are best presented on a small-area map in which more detail can be shown. ArcMap provides three types of charts: a pie chart, a bar or column chart, and a stacked chart. For each one, you will set the symbology method to Chart and select which fields to display in the chart. The map reader will use these fields for visual comparison, so the fields must offer some basis for comparison.

2-1
2-2
2-3
2-4

For pie charts, the fields selected must represent the whole of what the values represent. For instance, if you are representing marital status, you cannot show only the married and divorced. There are more categories, and you will not be representing all the people surveyed. Include categories such as single and widowed to represent all the people. By looking at the pie slices, the viewer can get a feel for the portion of the total that each value represents. Still, these slices do not lend themselves to showing values effectively. By looking at two pie charts, you may see that one category has a larger pie slice but you cannot tell the exact values.

Bar charts show actual amounts rather than the percentage of the total. Each value is shown as a bar in the chart, and a comparison can be made between the heights of the bars. Stacked charts also show values but display the columns on top of each other rather than side by side. The column heights are additive, with the total column height representing the total of all the field values.

Which chart style you use depends on whether you want to show percentages (pie charts) or compare actual values (bar or stacked charts).

Scenario   Continuing with the investment opportunity in ethnic-themed grocery stores, you have found an area that seems ripe for expansion. The bankers are considering your loan, but they want to make sure that you will have a chance at long-term success. It will be better if the houses in the area are primarily owner occupied so that each store can build a relationship with the customers and keep them coming back for a long time. If the houses in the area are mostly rental units, the customers may be more transient and not support the community efforts you are making. They may also not have as strong a customer loyalty as permanent residents. You also want to note the number of vacant houses, which may represent the opportunity for future customers who will take pride in their neighborhood.

Data   Continue using 2010 Census block-group-level data. A series of fields in the attribute table contain the occupancy status of the houses in the area, which you will be able to chart. Because the total of vacant houses, owner-occupied houses, and rental houses represents the total number of houses, you can use pie charts to show the data.

Examine the attribute table of the census data

1   In ArcMap, open Tutorial 2-4.mxd.

The map shows an area of great Hispanic concentration. Some ethnic-themed grocery stores are located near here, but the potential for expansion seems high. Now if you can prove to the bank that the housing population here is stable, it will be no problem to get the loan.

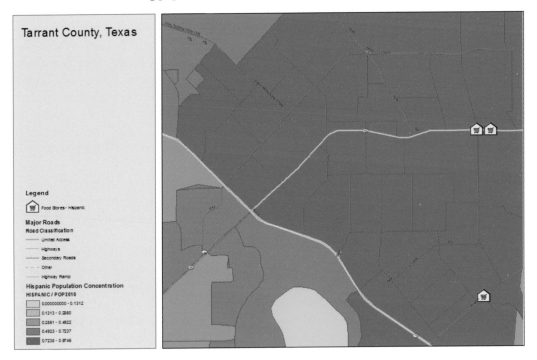

2   Right-click the layer Census2010 and open the attribute table.

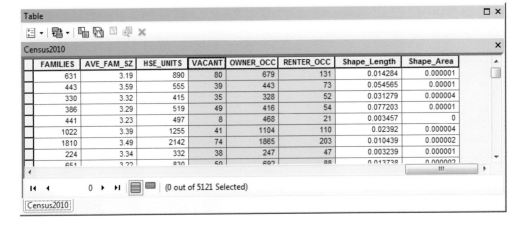

Notice the fields VACANT, OWNER_OCC, and RENTER_OCC. For these fields to be valid in a pie chart, the total of the values for each census block must represent the total number of housing units. To check the totals, choose one of the rows and add up the values for these three fields. They should equal the value of the field HSE_UNITS.

3 Close the attribute table and turn on the Census2010 layer.

Create a pie chart

1 Open the Properties dialog box for the Census2010 layer. On the General tab, change the layer name to **Housing Status**.

2 Go to the Symbology tab and click Charts > Pie in the Show window. Highlight the three fields for the chart—VACANT, OWNER_OCC, and RENTER_OCC—and click the > button. You can add these fields one at a time, or press and hold Ctrl and select them all together.

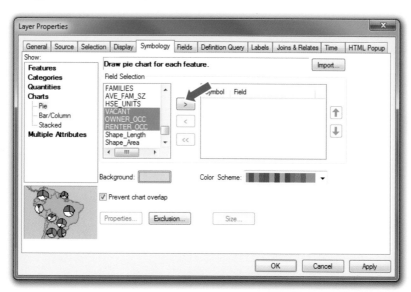

Remember, the fields you select must represent the whole; and in this case, they total the number of housing units. Once the fields are transferred, you will set the colors for each pie segment. The Hispanic concentration layer you already have uses darker hued colors, so you will select brighter colors for contrast.

3 Right-click the symbol box for VACANT and select Big Sky Blue from the color palette.

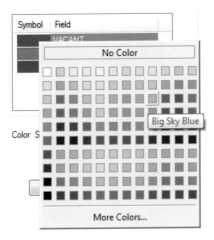

4 Right-click the symbol box for OWNER_OCC and select Quetzel Green. Change the symbol color for RENTER_OCC to Aster Purple. Click Apply.

The charts are now drawn, but the Housing Status layer covers the Hispanic Population Concentration layer. To correct the visibility problem, you will set the background of the Census2010 layer to clear.

5   Click the color box next to Background and select Hollow. Click OK and then OK again.

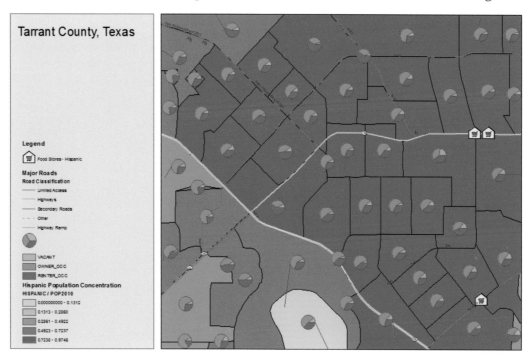

The map now displays the pie chart of housing status on top of the Hispanic concentration layer. By looking at the pie charts, you can see that the vacancy rate is low, and that most of the census blocks have a high percentage of owner-occupied housing. Only a few areas have a larger percentage of renter-occupied housing than owner-occupied housing, so you will be able to show the bankers that the neighborhood has a stable customer base.

YOUR TURN

Experiment with other types of charts and map scales.

- Change the chart type to Bar/Column and observe the differences. You may also want to try a stacked chart to see what effect it has on the map display. You may need to change the size to make the charts readable. When you are finished, change the chart type back to Pie Charts.
- Click Bookmarks > Zoom 2. What happens to the charts at this scale?

6   Save your map document as **Tutorial 2-4.mxd** to the \GIST2\MyExercises folder. If you are not continuing to the exercise, exit ArcMap.

Exercise 2-4

So far, you have shown an area of great interest for grocery store development by finding an area with a high Hispanic population. Then you demonstrated to the bankers that the housing status of the area is favorable—with a high percentage of owner-occupied homes.

In this exercise, you will repeat the process of creating a chart to show the male versus female population for each census tract.

- Continue using the map document you created in this tutorial, or open Tutorial 2-4.mxd from the \GIST2\Maps folder.
- Display chart symbols for male and female population; choose which style you feel will best display these values.
- Set appropriate colors for the chart.
- Change elements such as the titles, colors, and legend to make a visually pleasing map.
- Save the results as **Exercise 2-4.mxd** to the \GIST2\MyExercises folder.

What to turn in

If you are working in a classroom setting with an instructor, you may be required to submit the maps you created in tutorial 2-4.

Turn in a printed map or screen capture of the following:

Tutorial 2-4.mxd
Exercise 2-4.mxd

Tutorial 2-4 review

You looked at several types of charts that you can use ArcMap to create. The pie chart shows each value as a percentage of the whole—that is, if all the component fields equal the whole. It does not, however, give you a good feeling for amounts.

Bar charts and stacked charts give a better display of actual field values. Bar charts allow you to compare field values to each other. Stacked charts allow you to compare the totals for an area with the totals for other areas by looking at the height of the stack. However, both types of charts present other challenges in displaying the data so that it is easy to read.

Charts in general are an excellent way to show attribute data overlaid on other data. By setting the background of the chart layer to hollow, you can see underlying analysis displayed as well as the information from the charts. You must be aware of scale as charts tend to clutter and overlap on small-area maps. Another caveat with charts is not to include too many fields. The tutorial's pie chart with three fields was easy to read, but imagine how difficult it might be to read a chart with 14 fields. A bar chart with 14 fields tends to be wide and may not fit inside the polygon it represents.

2-1
2-2
2-3
2-4

Study question

1. What are the advantages and disadvantages of each chart style in regard to the following issues:
   - Scale
   - Values versus percentages
   - Number of fields
   - Other data layers

Other real-world examples

The police department has crime statistics for each police district in the city and wants to show a comparison of last year's crime rate versus this year's crime rate. A map that has the police districts color-shaded in the background might have a bar chart shown in each district that has both years' crime rates. You can compare the values by looking at the chart, and the activity in one district can be compared to another district by comparing the size of the bars.

The city secretary has compiled the election results and wants to display the vote totals for each precinct. She might create a map that has the voting precincts in the background, and a pie chart that shows each candidate's total represented by pie slice.

The National Energy Commission wants to display data to show how much of each state's energy comes from oil, natural gas, and nuclear power. It might create a map of the United States with a bar chart over each state to display the totals. The sizes of the bars can also be compared between states to determine which state uses the most energy.

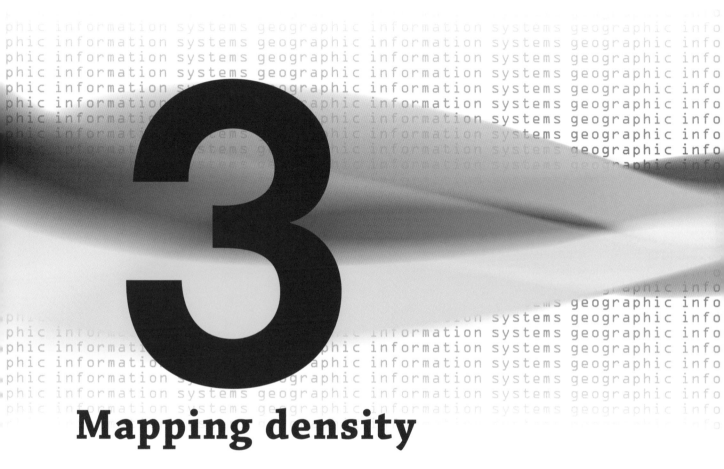

3

Mapping density

Values associated with an area can be shown simply as a value, but they may have a greater impact if shown in comparison with the area of a feature. The relationship between polygon feature values and the area of the polygon is called *density*. Using density to factor in the impact of area makes comparisons across features more precise. In most cases, no special symbology is required to show density. A simple color ramp from light to dark, with dark showing the highest density, is often sufficient. Other instances may use a special symbology called a *dot density* or a method of showing density with raster datasets called a *density surface*.

Tutorial 3-1

Displaying density for analysis

Data summarized by area, such as census data, is often displayed as a straight value or as a percentage of the total, as demonstrated in chapter 2. These values have no relationship to the size of the area they represent. Comparison across polygons of different sizes is difficult until you factor in area, creating a density value.

Learning objectives

- *Create density values*
- *Compare data across polygons of different sizes*

Preparation

- *Read pages 69-75 in* The Esri Guide to GIS Analysis, *volume 1.*

Introduction

In chapter 2, you looked at data shown by the amount contained in an area. It is easy to compare values across polygons as percentages or totals, but the values may give a false picture of the data.

An amount contained in a large area does not have the same significance as an equal amount contained in a small area. Imagine if you compare 50 people on one hundred acres of land to 50 people on one acre of land. Using density to display people per acre (amount divided by area) tells a different story from the count alone, especially if the polygons vary greatly in size.

Any value divided by the measured area it represents is called a density. It is common to hear the terms *people per square mile*, *value per square foot*, or *crop yield per acre*. Each of these amounts is a value divided by an area measurement. In fact, you can substitute the words "divided by" when you hear the word "per." Note that the area units can be different for each density calculation, so it is important to display the units of area on your map.

When density is shown on a map, area is removed as a factor in comparing values. A large residential subdivision with three housing units per acre is no denser than a small subdivision with three housing units per acre.

Scenario   The city planner wants to see a map of population totals and population density for the year 2010 in people per square mile. The first map is simply the population value field symbolized with graduated colors. The second map will divide the value by the area to display density, or people per area value. "Per" can mean "divided by" for your purposes, and if you recall, the word *normalize* also means "divided by." You should be able to use the normalization field in ArcMap to easily create a density map.

Data   The data is the 2010 Census block-group-level data. This particular dataset comes from the Data and Maps for ArcGIS data that comes with ArcGIS software. It contains fields for the total population in 2010 and 2012. The area needed is cut out for you to use, but the analysis can be repeated for any area with the data from Data and Maps.

The other set of data is the street network data to give the map some context. It also comes from Data and Maps.

3-1
3-2
3-3

Map population density by census block

1   In ArcMap, open Tutorial 3-1.mxd.

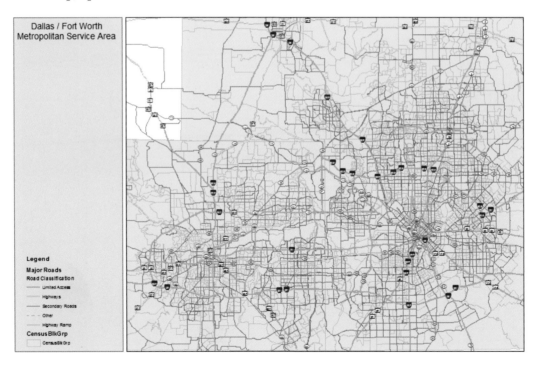

The map shows the Dallas/Fort Worth metropolitan service area, a four-county region with a population of about five million. You are interested in seeing the total 2010 population by census block group, and a population density of the same data. First, you will make a copy of the CensusBlkGrp layer, and then you will symbolize the two layers to display the desired data.

2   Right-click the CensusBlkGrp layer and click Copy.

3 Right-click the Tarrant County data frame and click Paste Layer(s). Turn the copied layer off for now.

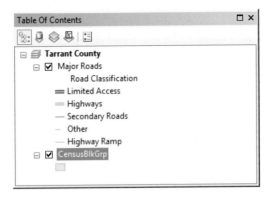

Now change the layer name and symbology properties for the first layer. This information is what will be displayed in the legend, so it should be descriptive of the data it represents.

4 Right-click the visible CensusBlkGrp layer and click Properties. Go to the General tab and type **2010 Population** for the layer name. Click Apply.

5 Go to the Symbology tab. In the Show box, click Quantities > Graduated colors.

6 Set the Value field to Population 2010. Accept the default classification and colors. Click Apply.

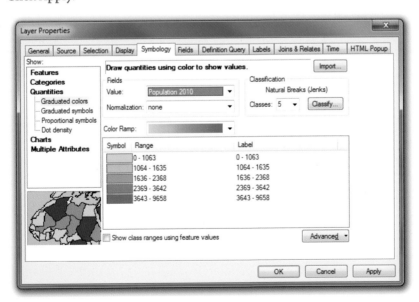

Examine the change in the map display. This map is a quick and simple display of population. The contrast looks good, but seeing the population count numbers in the legend may not mean a lot when viewed at this scale. You only want to know the relative population, so you will change the labels to Low, Medium, and High.

7 Click in the Label column for the lowest value and replace it with **Low**. Click the next label and press Delete to remove the label altogether. Make the middle label **Medium**, leave the next label blank, and name the highest value label **High**. Click OK.

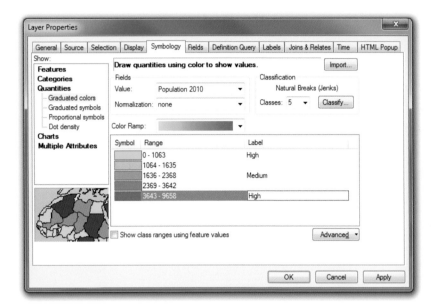

The map now displays the total population for each census block group, and your simplified legend makes the map easier to read. Note where the darkest areas are, and later you will see if these spots are, in fact, the areas of the highest population density.

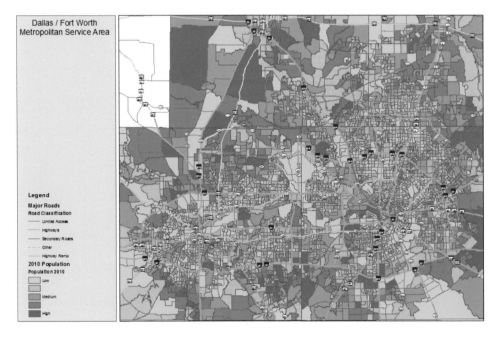

8 Print this map, or create an image file of it.

Next, you want to use the copy of the CensusBlkGrp layer to show population density, or people per area. You will use Population 2010 as the Value field again, and you can use the normalization function to automatically divide by area. The city planner wants to see the results in people per square mile, so you will need a field that contains the square mileage for each census block group. First, add a field to the table, and then you will calculate its values in square miles.

Map population density per square mile

1 Right-click the CensusBlkGrp layer and open the attribute table. Click the Table Options down arrow and click Add Field.

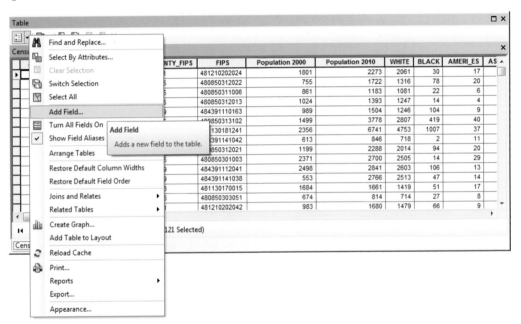

	NTY_FIPS	FIPS	Population 2000	Population 2010	WHITE	BLACK	AMERI_ES	AS
	1	481210202024	1801	2273	2061	30	17	
	5	480850312022	755	1722	1316	78	20	
	5	480850311006	861	1183	1081	22	6	
	5	480850312013	1024	1393	1247	14	4	
	9	484391110163	989	1504	1246	104	9	
		480850313102		1499	3778	2807	419	40
		130181241	2356	6741	4753	1007	37	
		391141042	613	846	718	2	11	
		850312021	1199	2288	2014	94	20	
	5	480850301003	2371	2700	2505	14	29	
	9	484391112041	2498	2841	2603	106	13	
	9	484391141038	553	2766	2513	47	14	
	3	481130170015	1684	1661	1419	51	17	
	5	480850303051	674	814	714	27	8	
	1	481210202042	983	1680	1479	66	9	

Table Options menu:
- Find and Replace...
- Select By Attributes...
- Clear Selection
- Switch Selection
- Select All
- Add Field...
  - Add Field: Adds a new field to the table.
- Turn All Fields On
- ✓ Show Field Aliases
- Arrange Tables
- Restore Default Column Widths
- Restore Default Field Order
- Joins and Relates ▶
- Related Tables ▶
- Create Graph...
- Add Table to Layout
- Reload Cache
- Print...
- Reports ▶
- Export...
- Appearance...

(121 Selected)

2 Enter **SqMiles** as the new field name, and set its type to Float in the Type drop-down list. Then click OK.

Add Field

Name: SqMiles

Type: Float

Field Properties

Alias	
Allow NULL Values	Yes
Default Value	

OK   Cancel

3 Right-click the new SqMiles field and click Calculate Geometry.

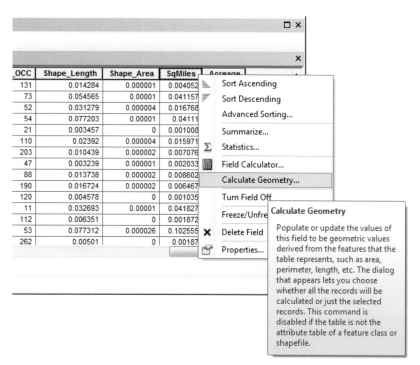

4 Click Yes on the warning message. In the Calculate Geometry dialog box, make sure that Property is set to Area, and use the drop-down list to set the Units field to Square Miles US. Then click OK.

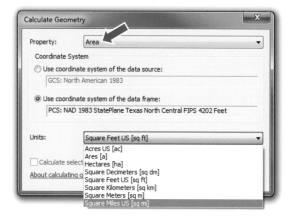

5 Click Yes on the warning message and close the attribute table.

6 Turn off the 2010 Population layer. Turn on the CensusBlkGrp layer, open its properties, and go to the General tab. Change the name to **2010 Population Density** and click Apply.

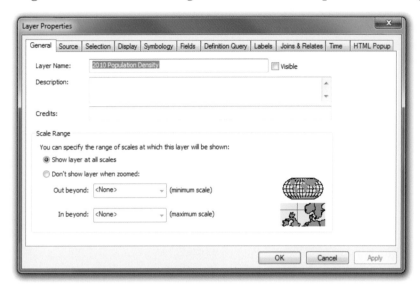

7 Go to the Symbology tab and click Quantities > Graduated colors. Set the Value field to Population 2010 and the Normalization field to SqMiles.

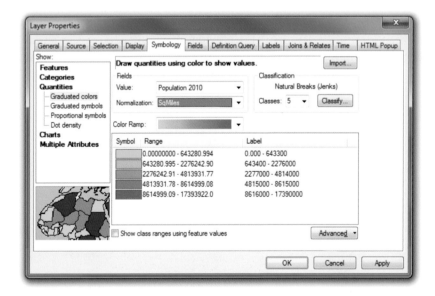

Once again you can simplify the legend so that the viewer gets a general feel for the data, and not necessarily concentrate on individual values.

8 Set the labels to match the 2010 Population layer, changing them to **Low**, **Medium**, and **High**. When everything matches the graphic, click OK.

3-1

3-2

3-3

This process adds a field, calculates an area, and uses that area to set up a display of density. Changing the labels in the legend makes the map easier to read and suitable for a general audience.

The resulting map is very different from the previous map. What you saw as large, dark polygons representing lots of people turned out to be medium-to-low density. Now it becomes easy to see the concentrations of people.

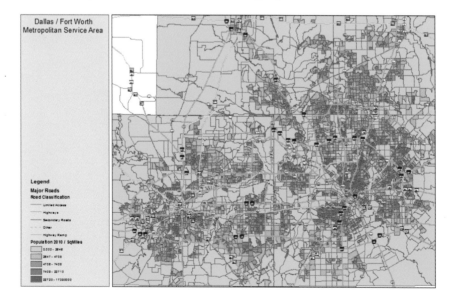

9 Print this map, or create an image file of it and save to the \GIST2\MyExercises folder.

YOUR TURN

Create two more maps: one that shows the 2000 total population and one that shows the 2000 population per square mile, using the Population 2000 field in the CensusBlkGrp layer. These maps will give the city planner two sets of data to compare, and to see how the population and the population density in the area have changed over time. Save these maps to the \GIST2\MyExercises folder.

These printed maps nicely display how things have changed from 2000 to 2010, but it may be hard to compare two specific areas. Flipping one sheet over the other is not efficient. Now look at an ArcMap tool on the Effects toolbar that will allow you to compare the two with ease.

Compare maps using the Swipe tool

1   Turn on the layers 2000 Population and 2010 Population, and turn off the two population density layers. On the main menu, click Customize > Toolbars > Effects.

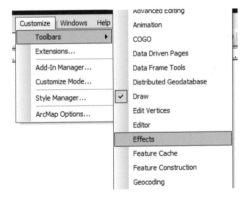

2   Because Effects tools work only in data view, on the main menu click View > Data View.

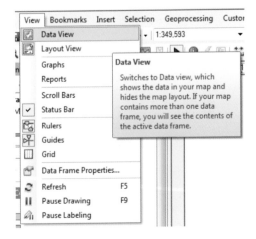

3 Make sure that the 2000 Population layer is above the
2010 Population layer in the table of contents by dragging
it higher in the list.

4 Set the layer in thae Effects toolbar to 2000 Population.

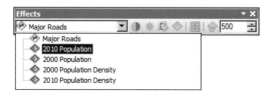

5 Click the Swipe tool 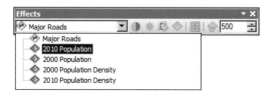 on the Effects toolbar.

6 Move into the right center of the map area, and then press and hold the left mouse button. It may take 10 or 20 seconds for the map to load into the Swipe tool memory. When the hourglass disappears, move the cursor to the left and notice that the 2000 Population layer is being replaced as you move the tool across the map.

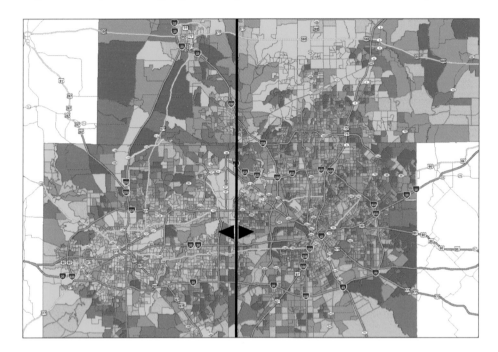

The Swipe tool is effective for seeing how features have changed between two layers. Grouped layers can also be set as a swipe layer, allowing you to swipe across sets of data.

YOUR TURN

Turn off the two population layers and turn on the two population density layers. Set up the Swipe tool for these layers and examine the changes over the 10-year span.

7 Save your map document as **Tutorial 3-1.mxd** to the \GIST2\MyExercises folder. If you are not continuing to the exercise, exit ArcMap.

Exercise 3-1

The tutorial showed how to map census values as both totals and densities. The values used were population counts, but other data is also useful when mapped as a density.

The city planner also wants to show the density of housing units, expressed in households per acre. The field HOUSEHOLDS in the census data has the number of households in each census block group.

- Continue using the map document you created in the tutorial, or open Tutorial 3-1. mxd from the \GIST2\Maps folder.
- Copy the Population 2010 layer and rename it **Household Density**.
- Add a field to the attribute table and calculate the acreage for each polygon.
- Symbolize the households using the HOUSEHOLDS field.
- Use the Quantile classification with five classes. Normalize by the new acreage field.
- Change elements such as the titles, colors, and legend to make a visually pleasing map.
- Save the results as **Exercise 3-1.mxd** to the \GIST2\MyExercises folder.

3-1
3-2
3-3

What to turn in

If you are working in a classroom setting with an instructor, you may be required to submit the maps you created in tutorial 3-1.

Turn in a printed map or screen capture of the following:
> **The four maps in Tutorial 3-1.mxd**
> **Exercise 3-1.mxd**

Tutorial 3-1 review

In this tutorial, you saw how density displays data differently from straight counts or totals. By dividing the total value by the area it represents, densities allow you to make comparisons between dissimilar areas by showing the data as a concentration. Because the normalization feature allows you to specify a field to divide into the Value field, you can easily use any field to set up density displays.

Although fields that show counts or totals summarized for an area benefit from display as a density, other fields may not be as suitable. Fields that display calculated values, such as averages or percentages, are not suitable for density maps. These fields are already divided by another value, so dividing by area does not make sense. A field that represents a percentage is the value divided by the total. Showing this percentage as a density is the value divided by the total, divided by the area. A map of "Percent Renter Occupied per Square Mile" has no real meaning.

Study questions

1. Can any field be used for density mapping?

2. Why do densities or concentrations allow a better comparison between values that are summarized by area?

Other real-world examples

A police department may summarize data by police beat and display the density of a particular crime. This type of crime mapping can help it determine how many officers to assign to each beat.

A political campaign might take voter registration rolls and summarize them by county. Then it can display the areas of concentration of registered voters to target its efforts.

Census data is one of the most common things to display as densities. Any of the data fields that represent totals or counts, such as people per square mile or concentrations of high school graduates, can be displayed as a density.

Tutorial 3-2

Creating dot density maps

One drawback of density maps that use shaded polygons or a density surface is that only one value can be mapped at a time. However, the dot density map allows you to display the density using a pattern that can be overlaid with other data.

Learning objectives

- *Create a dot density map*
- *Overlay density data*
- *Analyze patterns*

Preparation

- *Read pages 76–77 in* The Esri Guide to GIS Analysis, *volume 1.*

Introduction

In tutorial 3-1, you displayed data divided by the area it represents. This type of density mapping works well in most situations, but it causes problems if you want to display multiple values. The solution is to use a dot density map.

To create a dot density map, select a value field, a dot size, and a number of units that the dot will represent. ArcMap reads each value and calculates how many dots to display in the polygon area. For example, if the field value is 1,200 and the dot value is 20, ArcMap randomly displays 60 dots in the polygon.

It is important to note that the dots are placed randomly. This random placement works well in small areas, but in larger areas the dots may be randomly grouped in a region of the polygon, which suggests that the dots represent some sort of data grouping.

Another important factor in making a dot density map is the dot size. The size is set in conjunction with the dot value to determine how large the dots should be. Too large, and the dots will overlap and obscure the map; too small, and they will not be noticeable against the background. You may need to experiment with the dot size and dot value to get the best possible display for the scale of map you are producing.

Scenario   The parks director wants to build a dog park in Oleander and needs help in finding the right place. He feels that the park will be used most by apartment dwellers who do not have a large yard in which the dogs can play. So you will look for an area with a large population density as well as a concentration of rental units.

Data The data is the 2010 Census block-group-level data. This particular dataset comes from the Data and Maps for ArcGIS data that comes with ArcGIS. It contains the field Population 2010, which represents the population count in 2010, and RENTER_OCC, which represents the number of occupied rental units. The area needed is cut out for you to use, but the analysis can be repeated for any area with the data from Data and Maps.

The other set of data is the street network data to give the map some reference. It also comes from Data and Maps.

Map the density of rental units

1 In ArcMap, open Tutorial 3-2.mxd.

The map shows the City of Oleander with the 2010 Population Density already color-shaded. You will add the concentration of rental units to the population data. To keep the rental unit data from obscuring the population data, you will set a dot density classification and show the patterns on top of the population layer.

2   Right-click the CensusBlkGrp layer, open the properties, and go to the General tab. Type **Concentration of Rental Units** for the title and select the Visible check box to make the layer visible. Click Apply.

3   Go to the Symbology tab and set the Show window to Quantities > Dot density.

Notice that the dialog box will let you add multiple fields to the Symbol list for dot density. As with making charts, the list should contain values that can be compared on the same scale. You can use as many fields as you like, but too many fields will make the map difficult to interpret. Having more than one symbol value also adds a degree of difficulty to setting the dot size and scale. There will be more dots per polygon, and they may begin to interfere with each other.

4 Select the RENTER_OCC field and click the > button to add it to the Symbol list. Click
Apply and note the change in the map. Try different colors, dot sizes, and dot values
to get the map to look the way you want it. Click Apply to check the results of your
settings. When you are satisfied, click OK.

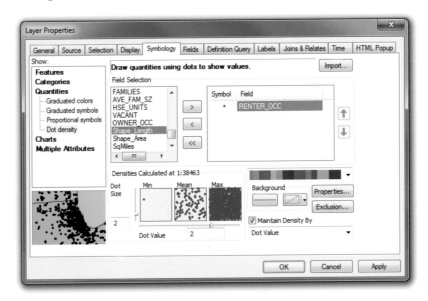

The resulting map shows the concentration of rental units as dot patterns, with the
underlying population density still visible. Click the Refresh View button ⟳ at the bottom
of your layout a few times and notice what happens to the dot distribution—it changes
each time. A quick visual analysis shows a place in the northeastern part of town that will
be perfect. It has a high population density and a concentration of rental units, as well as
being on a major thoroughfare. These characteristics are enough to investigate availability
of land and perhaps do a survey of what amenities might be desirable.

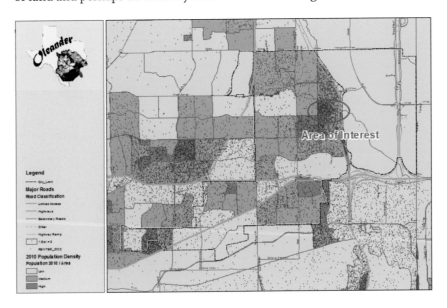

Controlling the location of the dots

There are two options for the dot density classification that can be used to add more control to the location of the dots. You can access these options by clicking the Properties button on the Symbology tab. The first option will lock the dots in place and keep them from moving around when you refresh the map. Although their placement is still random, it will allow you to keep a consistent look of the map for such things as a map series.

The second option can display the dots in relation to another polygon feature class. The first choice for restricting the dots keeps them from being displayed within the noted polygons. This choice may be useful if there are large lakes or ponds in the area. You do not want to display a density of people over a lake. The second choice displays the polygons in only the referenced area. Perhaps you want to show data from the census layer, but only within your city limits. Note that the restricting layer does not have to be the one for which you set the dot density classification.

5   Save your map document as **Tutorial 3-2.mxd** to the \GIST2\MyExercises folder. If you are not continuing to the exercise, exit ArcMap.

Exercise 3-2

The tutorial showed how to set up a dot density and display the density data over another color-shaded dataset. For this exercise, you will create another dot density map showing two fields as values.

The parks director wants to see the male-to-female comparison in the area of interest. You will try to show that data as a dot density map with the MALE and FEMALE fields from the census data.

Continue using the map document you created in the tutorial, or open Tutorial 3-2. mxd from the \GIST2\Maps folder.

- Copy the Concentration of Rental Units layer, change the name, and make it visible.
- Zoom to the bookmark Area of Interest.
- Set the symbology to Dot Density, and use both the MALE and FEMALE fields.
- Change the dot color, dot size, and dot value to make the data read well.
- Change elements such as the titles, colors, and legend to make a visually pleasing map.
- Save the results as **Exercise 3-2.mxd** to the \GIST2\MyExercises folder.

What to turn in

If you are working in a classroom setting with an instructor, you may be required to submit the maps you created in tutorial 3-2.

Turn in a printed map or screen capture of the following:
> **Tutorial 3-2.mxd**
> **Exercise 3-2.mxd**

Tutorial 3-2 review

In this tutorial, you saw how density can be displayed as dots. You were able to control the size and number of dots displayed. ArcMap, however, determines the actual random placement of the dots. Because of this feature, use the patterns the dots create as only a general view of density and not read them as grouping or clustering. Every time you redraw the map, the random location of the dots changes, potentially giving a different reading of the map. The dots can be locked down, but remember that they do not represent groupings, only a random scattering of dots.

Dot density maps can be created with several value fields, but you saw how too many fields can make the map hard to read and interpret. Care must be taken to set the dot value, size, and color.

One advantage of using dot density is that you are able to display a set of data on top of another symbolized layer. In this case, the dots were displayed over total population density so that the viewer can look at both the dot patterns and the color shading for visual analysis.

Study questions

1. When is it more practical to use dot density maps than charts?

2. Is dot density a true density?

3. A dot density map of the United States is created using each state as one polygon, and the total population as the value field. What is the problem with this scenario? What might the map look like? Can the patterns lead to false conclusions?

Other real-world examples

A police department may want to show concentration of crime without revealing actual victim locations. It might summarize crime data by police beat, and then create a dot density map of the totals to show general concentrations. The random assignment of dots will not represent real crime locations, but only symbolize that crime occurred.

A political campaign might want to display the number of registered voters as a dot density on top of the total population of voting age per precinct. Areas with a high eligible population but low registration might be a good target area for recruitment.

Census data lends itself well to density mapping, and dot density allows the overlaying of several types of data on a single map. With the total population as the backdrop, interesting maps can be created to show housing statistics in relation to other demographic categories.

Tutorial 3-3

Creating a density surface

A density map was easy to create using the data summarized by area. The polygons make it easy to create solid shade or dots to represent density. However, if the available data is point data, it is not as easy to show density. A density surface will take the points and create a continuous set of values to display density in a raster format.

Learning objectives

- *Create density surfaces*
- *Work with raster data*

Preparation

- *Read pages 78–85 in* The Esri Guide to GIS Analysis, *volume 1.*

Introduction

Tutorial 3-2 worked with data that was already summarized by area. It was fairly easy to color-shade the polygons to represent density, or to make a dot density map. With data that is not summarized by area, you must create a density surface in order to see the data projected as a density. A density surface is a raster image created from points, making a set of discrete features into a continuous phenomenon dataset.

When calculating a density surface, you set several parameters. The first is cell size, which equates to pixel size in the output data. Cell size can determine how fine or coarse the patterns will appear. A small cell size will create a finer resolution raster image but will take longer to process. Conversely, a large cell size will create a coarser image but will process quickly. ArcMap will suggest values based on the area you are working with.

The second parameter that you will deal with is the search radius. When ArcMap creates the density surface, it splits your map extent into pixels of the size you indicate. Then it counts all the features that fall within the search radius of each feature, divides that total by the area of that feature's search area, and creates a pixel with that value. Then it moves on to the next pixel, working its way through the entire dataset. A large search radius takes longer to process, and will give the surface a more generalized look. Small variations in the data may be missed because too many features are falling within each search radius. A small search radius reflects more local variation, with fewer features that have an effect on each pixel.

A density surface may reflect only the features and how they are grouped into the search area, or it may include a weight factor. The weight factor will give some features a more important role in the calculation of the density. For instance, a density surface that is made using only business locations might show data grouping near locations where there are many businesses. But if the data is weighted by the number of employees, the surface will show groupings nearer to businesses with many employees.

Scenario   The parks director was impressed with your analysis of suitable sites for the dog park. So he has come to you again with another request. Oleander is recognized as "Tree City USA" by the National Arbor Day Foundation, and the director wants you to do a tree inventory and tree density map of all the city parks. An intern with a Global Positioning System (GPS) unit spent all last summer mapping trees, and you now have the data she collected to make a tree density map. Because you are starting with point data, you will make a density surface to demonstrate tree concentrations. You will also use the measured canopy size as a weight factor, making the larger trees have a stronger effect on the calculation.

Data   The land-use data used to locate all the city parks is the city's cadastre file derived from recorded plat surveys. Each parcel of land has a land-use code, as you may recall from exercise 1-1.

The tree data was collected by an intern from Texas A&M University's School of Forestry. She visited every park in the city and recorded the longitude-latitude of every tree. She also built a data collection menu to collect data such as tree type, diameter, various condition measurements, and estimated canopy width.

The other set of data is the street network data to give the map some context. It comes from the Data and Maps for ArcGIS data.

Map tree density

1 In ArcMap, open Tutorial 3-3.mxd.

The map is zoomed in to the central part of the city, where a tree survey has been completed. The areas outlined in green are city parks, and the brown dots are tree locations. You may think that this point data reads well as density, similar to a dot density map, but there are a few things you will get from creating a density surface. First, you will get an output file that represents area, not discrete points. Plus, you will be able to weight the impact of the trees on the density calculation based on their canopy size.

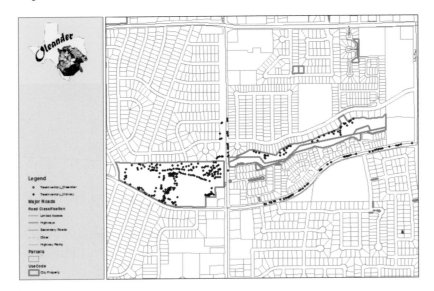

This tutorial uses the ArcGIS Spatial Analyst extension. You will activate the Spatial Analyst extension in this ArcMap session. If you are unsure whether you have this extension, click Customize > Extensions on the main menu and look for Spatial Analyst. Select it if it is not already checked. If it does not appear, ask your system administrator for access.

2 Click the Search tab on the right side of the map document window and type
point density in the Search window. Then click the Search button to find the tool.
Pausing over a tool name will show a description of the tool. Click the Point Density
tool to run it.

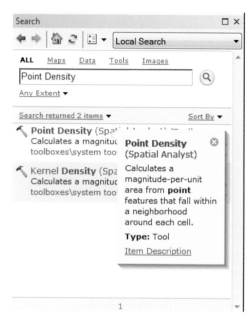

3 In the Point Density dialog box, set Input point features to TreeInventory_Midway
and the Output raster to Midway_Trees in the \MyExercises\MyData.gdb geodatabase.
Note: the raster output name is limited to 13 characters with no spaces.

The next values you will set are the search radius and the cell size. The Point Density tool
has already analyzed your data and map extent and suggests a suitable set of numbers. To
get a finer result, you can decrease these values by about 20 percent; and to get a coarser
result, increase them by the same factor.

4 Click the input box for Radius and type **53**. Replace the Output cell size with **6**. When your dialog box matches the graphic, click OK.

5 Look at the tree locations overlaid on the new density surface to get an idea of how the two sets of data relate. When you are finished, turn off the TreeInventory_Midway layer to get the final map.

The parks director can show this map to the park board and help it plan the tree planting efforts for next year. You can also use this analysis for area overlays in future GIS analysis, such as showing the effect on the existing tree density of future playgrounds or athletic fields.

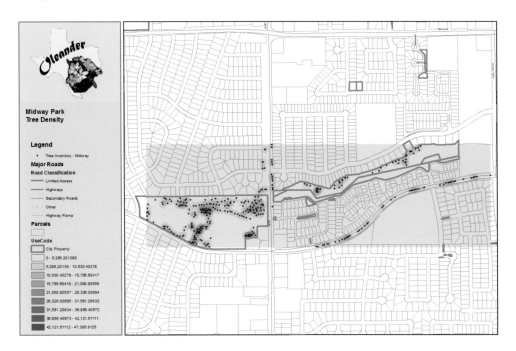

3-1
3-2
3-3

YOUR TURN

Move to the bookmark Oleander Park and repeat the density calculation for the TreeInventory _Oleander layer. Eventually, the director will want this calculation done for every city park.

6 Save your map document as **Tutorial 3-3.mxd** to the \GIST2\MyExercises folder. If you are not continuing to the exercise, exit ArcMap.

Exercise 3-3

The tutorial showed how to map densities from point data. The points were used to create a density surface, and the result was a continuous phenomenon dataset.

Now the parks director wants you to do a more detailed study of tree density. He wants to see the density of elms and crape myrtles on two more maps.

- Continue using the map document you created in the tutorial, or open Tutorial 3-3. mxd from the \GIST2\Maps folder.
- Return to the Midway Park bookmark. Open the properties of the TreeInventory_ Midway layer and set up a definition query to show only the elms.
- Create a point density raster of the resulting set.
- Change the definition query to **crape myrtle** and create a point density raster.
- Make maps of each of these point density rasters.
- Change elements such as the titles, colors, and legend to make visually pleasing maps.
- Print or export images of the two maps and save to the \GIST2\MyExercises folder.
- Save the results as **Exercise 3-3.mxd** to the \GIST2\MyExercises folder.

What to turn in

If you are working in a classroom setting with an instructor, you may be required to submit the maps you created in tutorial 3-3.

Turn in a printed map or screen capture of the following:

The two map layers you created in Tutorial 3-3.mxd
The two map layers you created in Exercise 3-3.mxd

Tutorial 3-3 review

The Point Density tool is used to make a continuous phenomenon dataset from discrete features. The surface can be used for further analysis with the data representing the values interpolated across a large area.

The tool creates a raster dataset with the specified cell size, and then evaluates each cell for features that fall in and around it. Once each cell is given a value, a colored raster image displays densities or concentrations.

Care must be taken to set the cell size and search area. Setting these values high will create a coarser map but process quickly. Setting them low will take more computer time but yield a finer result.

3-1
3-2
3-3

Study question

1. What are the three types of density mapping discussed in chapter 4 of *The Esri Guide to GIS Analysis*, volume 1, and when should each type be used?

Other real-world examples

The local weather reporter gets rainfall and temperature readings based on discrete measuring stations. The reporter then takes the point data and creates a density surface to show a continuous phenomenon of temperature and rainfall.

A fire department may take point locations of calls for service and create a density surface showing the concentration of calls.

A wildlife researcher might take point data of animal observations and create a surface density to show animal density. This process takes what is known to be limited sample data and interpolates it across a larger area.

4

Finding what's inside

GIS data deals with relationships among datasets, allowing for many analyses to be performed using these relationships. One of the most basic is the concept of data overlays. For example, does one set of data overlay another? Overlays might be used simply as a selection process, with analysis performed against the results. More complex processes might involve features that are only partially inside the overlay data. Analysis against the selection used in this situation cannot deal with the whole values for these features, but uses proportions instead.

Tutorial 4-1

Overlaying datasets for analysis

Finding features inside a region is a powerful analysis tool. The region can be a graphic drawn by hand, or a feature that exists in another dataset. Once the features are selected, summations and comparisons can be made using the attribute information.

Learning objectives

- *Create a visual overlay*
- *Select features*
- *Obtain field summaries*

Preparation

- *Read pages 87–104 in* The Esri Guide to GIS Analysis, *volume 1.*

Introduction

A powerful concept of spatial analysis is to decide whether one set of features is inside another set. For example, you might want to know if your house is in a floodplain, an earthquake zone, or a good school district. To find out using GIS, a polygon region must be defined, and then you can search for points, lines, or polygons that fall within those boundaries.

The boundaries used for an inside-outside analysis can be created as a graphic in ArcMap, or may exist as a feature in another dataset. The graphics are drawn using the standard drawing tools in the layout view and can be used to do simple selections.

Using features from another layer for your selection is a little more complex but a lot more powerful. The desired boundary must be selected, and a select-by-location query must be built. Conveniently, the query you build can be easily repeated for other regions. For example, to get a count of houses in a fire district, select one of the district boundaries from the Fire Districts feature class and use it in a select-by-location query against the residential parcel data. To repeat this process for another district, change the selected district and repeat the select-by-location query.

Once the features are selected, you can do many things with the associated data. You can get a count of the features. The preceding example is a simple answer to how many houses fall within a fire district. You may also want to perform a summary operation against the data. The summary can be the total value of houses or the average water usage.

You can also create boundaries by buffering features, but because buffering deals with setting a distance from an existing feature, it is covered in the "proximity analysis" tutorials in chapter 5.

Scenario   The city manager has identified a large area of land that represents the gateway to the city. The area has many shops, some vacant land, and an apartment complex. She has outlined the region on a map, and you will select the property within that region and provide some summary statistics, such as how many parcels and dwelling units, what the total area is, and how much of each of the land-use categories falls within the region.

Если the city council vote is favorable, the city will use its parks department staff and plants from the city's greenhouse to implement a beautification project in the area. The city council meets tonight, and it has asked for a briefing on the project.

Data   The dataset you will work with is the cadastre layer for Oleander. It represents platted property (property that is surveyed and recorded with the county land office) and unplatted property (property that does not have a survey recorded) and contains a field for land-use codes (UseCode). The field DU represents the number of dwelling units per parcel, and you will use that number for a housing count. You will also use the Shape_Area field, which ArcMap automatically provides, for the area calculations.

Select features using a polygon

1   In ArcMap, open Tutorial 4-1.mxd.

The map has a region outlined in red with a gray crosshatch. This area is where the city's parks department has an interest in developing more greenery and plantings as an entry point to the city.

Do a preliminary investigation by looking at the map. Basic visual analysis shows that inside the region there is some vacant land, some commercial uses, some multifamily housing, and some single-family dwellings. So the city manager already has a general idea of land use inside the region and can give the city council a briefing while you take more time to complete the work. This investigation is a simple but effective preliminary analysis.

To get the rest of the information, you will select the parcels that fall within the region. First, try doing a simple selection with a user-defined area.

2 Click the List By Selection tab at the top of the Table Of Contents window.

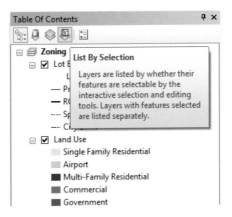

3 Click the selection icon to make Land Use the only selectable layer.

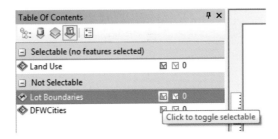

4 On the Tools toolbar, click the Select Features tool down arrow and click Select by Polygon.

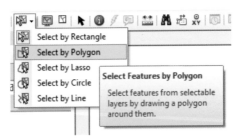

The Select by Polygon tool allows you to draw a freehand polygon around the area you want to select and then select whatever the selectable layers are. Always make sure to check what layers are selectable, or you can end up with a mess afterward.

5 Using the selection tool, draw a polygon that follows the outline of the shaded region. Click to insert a vertex. Double-click the last point to finish. If you make a mistake, click the Undo button ↺ on the Standard toolbar and start over.

The desired parcels are selected. You can now work with them to get the counts and summaries you want. This method of selection worked well, but there can be some problems with it.

6 On the main menu, click Selection > Clear Selected Features. Or click the Clear Selected Features button ▣ on the Tools toolbar.

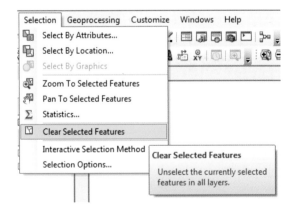

This selection method is not easy to repeat, and there lies one of the problems. Your task called for doing a summary, and now the selected set is cleared. Next, you will try another selection method. This method involves drawing a graphic of the area you want, and then using it to make the selection.

Select features using graphics

1   The Select By Graphics function works only in the Data View window. On the main menu, click View > Data View. Click the Select Elements tool on the Tools toolbar.

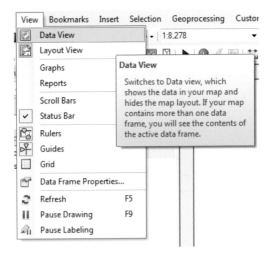

2   In the map area, select the shaded graphic that the city manager drew. It is hard to see on the map, but a rectangle that encompasses the shaded area is selected, as evidenced by a blue dotted line and selection points.

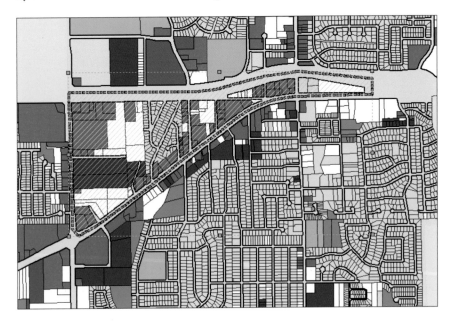

3 On the main menu, click Selection > Select By Graphics.

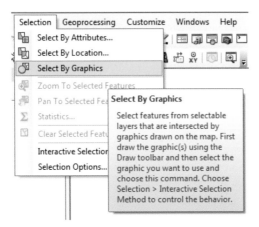

All the features inside the region of interest are again selected.

This graphic element surrounds the area of interest. So if the selected features are accidentally cleared again, you can quickly select them using this graphic. If there is no graphic element to use, you can draw one using the New Polygon tool on the Draw toolbar. These elements can be symbolized in the same way that regular data features are symbolized.

Now you can proceed with the counts and summaries that are required for this analysis task. The first step is to get a count of how many parcels are involved in the project.

Obtain counts and summaries for features within an area

1 Go to the List By Selection tab at the top of the table of contents. The number that appears to the right of the Land Use layer is the number of features that are currently selected.

It looks as if there are 202 parcels in the redevelopment region. That calculation was easy.

Next, you will determine how many dwelling units are in the region. The field DU has a value that represents the number of dwelling units on each parcel. Vacant and commercial lots have a value of 0, lots that contain single-family houses have a value of 1, and lots with multifamily dwellings reflect how many of them are on each parcel.

2   Open the attribute table of the Land Use layer from the Selection tab by right-clicking the layer and clicking Open Table Showing Selected Features.

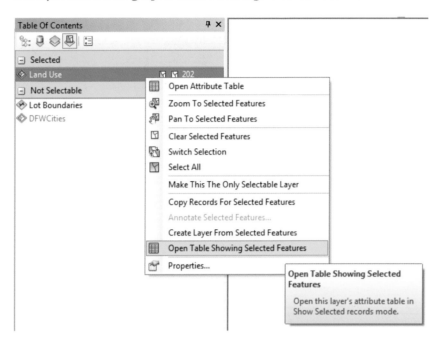

The table opens to display selected features. You can verify the number by checking the bottom of the table window. Yes, 202. Now get the total dwelling units.

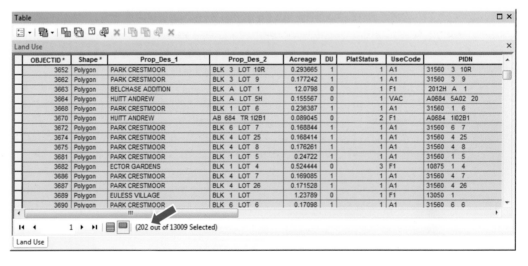

3   Right-click the field DU and click Statistics.

The resulting window displays the count, minimum, maximum, sum, mean, and standard deviation. Once again, you can verify the count at 202 and see that the total number of dwelling units is 633.

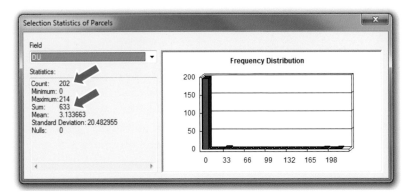

You will also get a total area of all the parcels. For total area, use the statistics feature again.

4   Close the Statistics window. Right-click the field Acreage and click Statistics.

The resulting window shows all the statistics of the field, including total acreage.

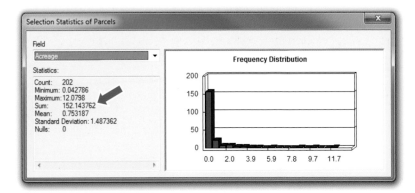

The total acreage is 152.1 acres. The city manager wants only one more piece of information: the breakdown of area for each land use. This analysis so far has produced only the overall total. To get the breakdown, you will use the Summarize command. This method will take the selected features and a selected field, and then report back each unique value of that field. In addition, you can add summary statistics to the process and get totals, averages, or standard deviations for the other fields in the table.

For this task, use the UseCode field. The Summarize command will list each unique value in the Land Use field, and allow you to add a command to get the total acreage for each category.

5   Close the Statistics window. Right-click the UseCode field and click Summarize.

6   In the Summarize dialog box, click the plus sign next to Acreage and select the Sum check box. Sum gives the total acreage for each unique value of UseCode in the table. If necessary, select the check box next to "Summarize on the selected records only." Click the Browse button next to the "Specify output table" box and store the table as **LandUseSummary** to the \GIST2\MyExercises\MyData.gdb folder. Click Save. When your dialog box matches the graphic, click OK to start the summary process.

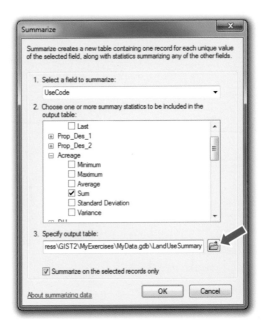

For more information on performing field summaries, click About Summarizing Data in the summary dialog box.

7 When the process is finished running, click Yes to add the table to the map document. Close the attribute table and click the List By Source button at the top of the table of contents. You will see the summary table that you created.

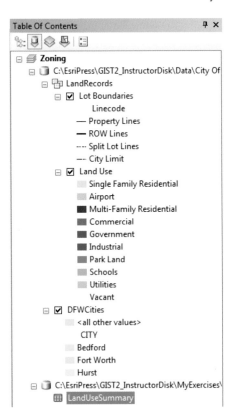

8 Right-click the table and click Open to view the results.

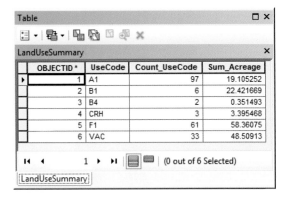

The table shows many of the results that the city manager wanted, and just in time for the meeting. Codes are defined as follows:

- A1   Single Family Residential
- B1   Multi-Family Residential
- B4   Duplex
- CRH   Church
- F1   Commercial
- VAC   Vacant

YOUR TURN

Switch to layout view and add a text box to the map that displays the total number of parcels, total number of dwelling units, total acreage, and the breakdown of acreage per land-use type.

This exhibit will be perfect to distribute at the city council meeting and get support for this project.

9 Save your map document as **Tutorial 4-1.mxd** to the \GIST2\MyExercises folder. If you are not continuing to the exercise, exit ArcMap.

Exercise 4-1

The tutorial showed how to select features inside a box in two ways: manually select and select using a graphic.

In this exercise, you will repeat the process using a different area of the map. There is another area at the north end of town that the city manager has circled with a graphic. The manager wants you to calculate the same statistics for that region as you did for the first one.

- Continue using the map document you created in the tutorial, or open Tutorial 4-1. mxd from the \GIST2\Maps folder.
- Go to the bookmark Site 2.
- Select the features inside the marked area using one of the techniques from the tutorial.
- Perform statistics or summaries to get the following:
  - Number of parcels
  - Number of dwelling units
  - Total acreage
  - Breakdown of acreage by land use:
    - B1 = Multi-Family Dwelling Units
    - F2 = Industrial
    - POS = Public Open Space (Park)
    - VAC = Vacant
- Add these values to your map layout.
- Change elements such as the titles, colors, and legend to make a visually pleasing map.
- Save the results as **Exercise 4-1.mxd** to the \GIST2\MyExercises folder.

What to turn in

If you are working in a classroom setting with an instructor, you may be required to submit the maps you created in tutorial 4-1.

Turn in a printed map or screen capture of the following:

Tutorial 4-1.mxd
Exercise 4-1.mxd

4-1

4-2

Tutorial 4-1 review

This tutorial showed two techniques to find what is inside a specified region. The regions were drawn using either a selection box or with a graphic that stays on the map. The first technique works well, but if you accidentally clear the selected features, you must start all over again. The second technique uses a graphic element for the selection. The element is more permanent and can be used over and over. It can also be symbolized and used as part of the map layout to mark the area of interest.

Once the features are selected, you can use two attribute table functions to get information about the data. The first gives you simple statistics for a selected field. The Statistics command returns the minimum, maximum, sum, count, mean, and standard deviation. The second command, the Summarize command, allows you to do more with a selected field. It returns every unique value in the selected field, and then allows you to perform any of the statistics operations against the other fields. Using this tool, you can obtain the sum of the acreage for each land-use type.

Study questions

1. What happens to features that cross the selection boundary?

2. Look at the summary operations for text fields. How do they differ from the operations for numeric fields?

3. Is every numeric field suitable for summary operations?

Other real-world examples

The police department might need to close a road for the Independence Day parade. The department can draw a boundary on the map and use it to select property owners to notify.

After noticing some clogged storm drain inlets, the public works department might draw a boundary around the clogged inlets and use the map to select pipes to inspect.

Tutorial 4-2

Finding features partially inside

Features that fall completely within the study area boundary are easy to work with, but sometimes the features cross a boundary line. These features must be handled a different way to extract correct data from them.

Learning objectives

- *Select features in relation to a boundary*
- *Understand different overlay functions*

Preparation

- *Read pages 87-104 in* The Esri Guide to GIS Analysis, *volume 1.*

Introduction

When using boundaries to select features, not all features will fall completely within the boundary; they may share an edge with the boundary, or they may cross the boundary. The attributes associated with the feature represent the whole feature. If only a portion of the feature is inside the boundary, only a portion of the attribute value can be used for overlay analysis.

As an example, examine some census data. Each census block has a count of how many people are in it. If a feature's boundary crosses through a census block, how many people will be shown to live inside the boundary? Not all of them—only some portion of them. How much square footage of the census block is inside the polygon? ArcMap calculates area automatically for each polygon feature (Shape_Area), so the census block must be split at the boundary to get the total. If the straight totals are used, the results will be wrong.

When the selection of features within a boundary returns whole polygons or lines, it is OK to get a count or summary as long as you know you are getting the values for the whole polygons. When features cross the boundary and you want values for only the areas inside the boundary, you can split the polygons at the boundary and, in some cases, proportionally divide the values. For census data, you can calculate the proportional population within the smaller units by multiplying the density times the area. It is important to note, however, that the resulting values are only a broad estimate. Splitting the values proportionally will not take into account any concentration of values within the polygon. Comparing the results to more accurate data will most certainly highlight the inaccuracies achieved with this method.

Scenario   The city manager took your map to the city council for review. The council member who represents that region remembered that there are some flood zones that can affect this property. The council member wants some of the same calculations performed again, but only for the areas inside the floodplain, including a breakdown of how much of each of the flood zones is in the target area. The city engineer expects this information in the morning so that the city can move forward on the project.

Data   The dataset you will work with is the cadastre layer for Oleander. It represents platted and unplatted property, and contains a field for land-use codes (UseCode). The field DU represents the number of dwelling units per parcel, and you will use that field for a housing count. You will also use the Shape_Area field, which ArcMap automatically provides, for the area calculations. Overlaid on the cadastre layer are the flood zones as determined by the Flood Insurance Rate Maps (FIRM) from the Federal Emergency Management Agency (FEMA). You will use the overlay layer to determine what is inside the floodplain. This layer also has the flood zone categories in the field Zone, as follows:

    A   100-year frequency storm (can be filled and reclaimed)
    B   50-year frequency storm (can be filled and reclaimed)
    FW Floodway (cannot be reclaimed)
    W   Standing water (lakes and ponds)

There is also a field named ZoneName that contains the name of the water drainage shed for each flood-prone area.

Select features by location

1 In ArcMap, open Tutorial 4-2.mxd.

The map is similar to tutorial 4-1, showing the land use and the area of interest. You also see the flood-prone areas. A quick visual inspection shows that some parcels are only partly inside the flood zone, so you will do calculations involving only those areas.

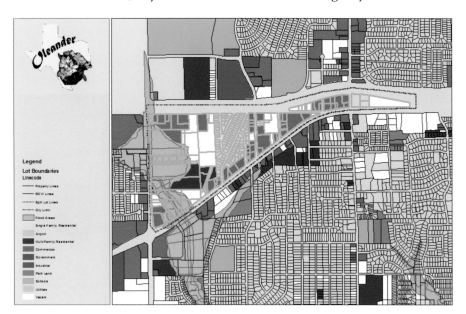

You will find out how many parcels are in the floodplain, what the total area of land is inside the floodplain, a breakdown of land use by area inside the floodplain, and a breakdown of how much of each flood zone category is in the selected area.

The 202 parcels inside the study area are already selected. If your selection is cleared, repeat the procedure from the previous tutorial and use the graphic's boundary to select them.

You will use the flood zone boundaries to select the parcels that intersect them using the Select By Location tool. This tool allows you to select features from one layer based on their spatial relationship to features in another layer. First, write down what you want to do:

I want to select features from the currently selected set of the Land Use layer that intersect the Flood Areas layer.

Now you will set the parameters in ArcMap to make the desired selection.

2 On the main menu, click Selection > Select By Location.

3 In the Select By Location dialog box, change the Selection method to "select from the currently selected features in."

4 In the Target layers(s) pane, select the Land Use check box.

If your list does not contain all the layers, clear the check box next to "Only show selectable layers in this list."

5 Set the Source layer to FloodAreas.

6 Set the Spatial selection method to intersect the source layer feature.

There are a lot of options to describe the relationship you want between the two layers. *Intersect* means that any portion of the parcels falls within the flood area. Some of the other options may mean that the features are totally inside the flood area, and some may mean that the features only touch the line that represents the border of the flood area.

7 When your dialog box matches the graphic, click OK to make the selection.

ArcMap makes a new selection from the previously selected features that shows which parcels intersect the flood area. Notice that some parcels cross the boundary, and some parcels are completely within the boundary of the Flood Areas layer.

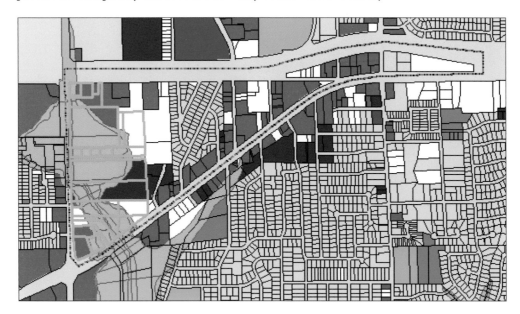

Some of the functions that you want to perform will not be affected by the whole parcel being selected. For instance, the first thing you want to know is how many parcels are affected.

8 Click the List By Selection button at the top of the table of contents. How many Land Use parcels are selected?

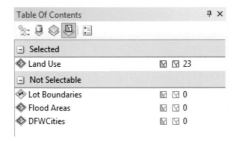

Your first statistic for the city manager is that 23 properties in the study area are inside the floodplain.

Next, you will find out how much acreage those properties represent. You cannot total the acreage because it will include the areas of some of the parcels that fall outside the floodplain. You will perform an overlay to cut the parcels at the flood boundary.

Create an overlay using selected features from different layers

1   Click the Search tab and search for the Overlay toolset. Pause your cursor over each of the results to see a description of the tools.

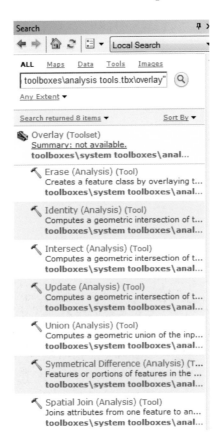

The tools that perform the type of overlay you want to perform are Intersect, Spatial Join, and Union. If you have an ArcGIS for Desktop Advanced license, you will have more choices in the Overlay toolset. You can read in ArcGIS for Desktop Help what the other tools do, but this tutorial uses Union.

Union takes the input feature classes and merges them, creating an output layer that contains the features split at their computed intersections. The result is that all the features of both layers are transferred to the output. In addition, all the attributes from both feature classes are preserved in the output.

Union (Analysis)

Title   Union (Analysis)

Summary
Computes a geometric union of the input features. All features and their attributes will be written to the output feature class.

Illustration

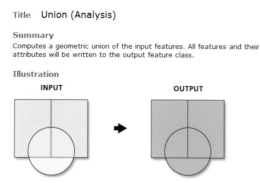

You will use the Land Use layer and the Flood Areas layer as the input layers.

2 In the Search results, click the Union tool to open its dialog box. Use the drop-down list and set Input Features to Flood Areas.

3 Use the drop-down list again to add Land Use to the Input Features list.

4 Click the Browse button next to "Output Feature Class," name this feature class **FloodAreaOfStudy** (do not include any spaces), and store it in LandUseReport.gdb in the \GIST2 \MyExercises folder.

5 When your dialog box matches the graphic, click OK.

The resulting feature class has the selected parcels that fall within the floodplain, as well as the floodplain areas from throughout the city. Some portions of the parcels that are not within the floodplain are still in this layer, and some portions of the floodplain that are not part of this analysis are still in the output layer. You can remove them using a definition query. ArcMap makes this function easy to do. Any parcel feature that does not intersect the flood zone layer has an FID_FloodZone value of –1, and any floodplain area that does not intersect a parcel has an FID_Parcels value of –1. The definition query will exclude features with an FID_FloodZone value of –1 and an FID_Parcels value of –1. The query includes the operator < >, which means "not equal to."

6 Switch back to List By Drawing Order in the table of contents and right-click the FloodAreaOfStudy layer. Then click Properties. Go to the Definition Query tab and click Query Builder. Build the query shown and verify it. Then click OK and OK again.

The FloodAreaOfStudy layer now shows only the areas that are within both the selected set of parcels and the floodplain.

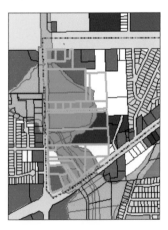

Next, you want to calculate acreage. The attribute table has a field named Acreage that you may be able to use.

Calculate acreage

1 Right-click the FloodAreaOfStudy layer and open the attribute table. Look for the Acreage field.

Prop_Des_2	Acreage	DU	PlatStatus	UseCode	PIDN	Prefix	StName	Suffix	SufD
BLK A LOT	8.99843	192	1	B1	8897 A	S	INDUSTRIAL	BLVD	
BLK A LOT	8.99843	192	1	B1	8897 A	S	INDUSTRIAL	BLVD	
BLK A LOT	8.99843	192	1	B1	8897 A	S	INDUSTRIAL	BLVD	
BLK A LOT	8.99843	192	1	B1	8897 A	S	INDUSTRIAL	BLVD	
BLK A LOT	8.99843	192	1	B1	8897 A	S	INDUSTRIAL	BLVD	
BLK A LOT	8.99843	192	1	B1	8897 A	S	INDUSTRIAL	BLVD	
BLK A LOT	8.99843	192	1	B1	8897 A	S	INDUSTRIAL	BLVD	
BLK B LOT ALL	9.91819	214	1	B1	44720 B ALL C		VILLA	DR	
BLK B LOT ALL	9.91819	214	1	B1	44720 B ALL C		VILLA	DR	
BLK B LOT ALL	9.91819	214	1	B1	44720 B ALL C		VILLA	DR	
BLK B LOT ALL	9.91819	214	1	B1	44720 B ALL C		VILLA	DR	
BLK B LOT ALL	9.91819	214	1	B1	44720 B ALL C		VILLA	DR	
BLK B LOT ALL	9.91819	214	1	B1	44720 B ALL C		VILLA	DR	
BLK B LOT ALL	9.91819	214	1	B1	44720 B ALL C		VILLA	DR	
BLK A LOT 1	12.0798	0	1	F1	2012H A 1	S	INDUSTRIAL	BLVD	
BLK A LOT 1	12.0798	0	1	F1	2012H A 1	S	INDUSTRIAL	BLVD	
BLK A LOT 1	12.0798	0	1	F1	2012H A 1	S	INDUSTRIAL	BLVD	

(0 out of 83 Selected)

FloodAreaOfStudy

The Acreage column holds values, but some values repeat. ArcMap did not automatically update the areas represented here; it only updated the field Shape_Area. You will recalculate the acreage value for this field.

2 Right-click the Acreage field and click Calculate Geometry. Click Yes on the warning. In the Calculate Geometry dialog box, select Acres US [ac] in the Units drop-down list and click OK.

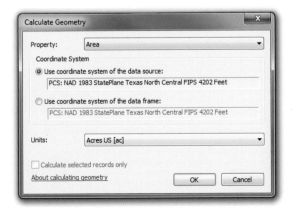

You are finally ready to do the area calculations for the Land Use types and the Flood Zone categories. The process will involve doing a field summary on both the UseCode field and the ZONE_ field.

3 Right-click the field UseCode and click Summarize. In the dialog box, expand the choices under Acreage and select the Sum check box.

4 Click the Browse button next to "Specify output table," name the table **StudyAreaUse**, and save it to the \GIST2 \MyExercises\LandUseReport.gdb folder. Click Save. When your dialog box matches the graphic, click OK. When prompted, click Yes to add the table to the current map document.

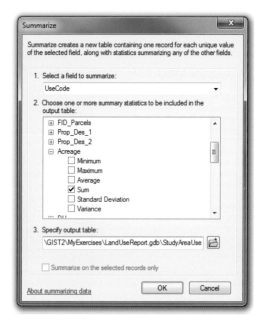

The results are recorded in a new table that includes the total acreage for each unique land-use code.

The city manager also wanted a similar summary operation to show acreage of each of the flood zone categories.

YOUR TURN

Repeat the summarize process using the ZONE field to get the total acreage for each flood zone area. Name the output file **StudyAreaZone** and add it to the map document. When you are finished, close the attribute table of the FloodAreaOfStudy layer.

5 On the List By Source pane of the table of contents, right-click StudyAreaUse to open the table.

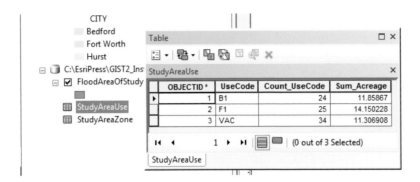

Here is the next set of data that the city manager wants. The acreage of each use-code value is summed in this table.

6 Right-click the StudyAreaZone file and open it.

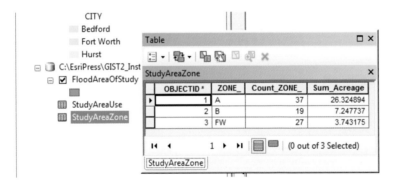

This table has the acreage for each flood zone category. Totaling the Sum_Acreage field will yield the total acreage inside the flood area—the final information you want to find.

7 When you have finished looking at the tables, close them. Add another text box to your layout and add all the information that the city manager wants.

8 Save your map document as **Tutorial 4-2.mxd** to the \GIST2\MyExercises folder. If you are not continuing to the exercise, exit ArcMap.

4-1

4-2

Exercise 4-2

The tutorial showed how to select features using features of another layer. In this exercise, you will repeat the process using a different area of the map. There is another area at the north end of town that the city manager has outlined with a graphic. The manager wants you to calculate the same statistics about that region as you did for the first one.

- Continue using the map document you created in the tutorial, or open Tutorial 4-2.mxd from the \GIST2\Maps folder.
- Go to the bookmark Site 2.
- Select the features inside both the study area and the flood zone using the techniques from the tutorial.
- Perform statistics or summaries to get the following:
  - Number of parcels
  - Total acreage
  - Breakdown of acreage inside the flood zone by land use
  - Breakdown of acreage inside the flood zone by flood zone category
- Add these values to your map layout.
- Change elements such as the titles, colors, and legend to make a visually pleasing map.
- Save the results as **Exercise 4-2.mxd** to the \GIST2\MyExercises folder.

What to turn in

If you are working in a classroom setting with an instructor, you may be required to submit the maps you created in tutorial 4-2.

Turn in a printed map or screen capture of the following:

Tutorial 4-2.mxd
Exercise 4-2.mxd

Tutorial 4-2 review

Not all boundaries used for analysis will be graphics or user-drawn boundaries. Many times they will be existing features. Using the Select By Location tool, you were able to use one set of features to select another set that had a spatial relationship. The relationship can be defined as inside the boundary, completely inside the boundary, crossing the boundary, or other options.

Merely selecting the features enabled you to do certain attribute calculations. You were able to get a count of features inside the boundary and make a list of the features. But because some of the features crossed the boundary, you were not able to get a total area. To get the total area, you split the polygons at the boundary to create new features, a process described in *The Esri Guide to GIS Analysis*, volume 1, as *feature overlay analysis*.

Study questions

1. What must be done to data summarized by area when its features are split at a boundary?

2. Can any feature type be used in a Select By Location operation?

3. Can any feature type be used with the feature overlay analysis tools?

Other real-world examples

A building department might make a map of all the inspections that are scheduled for the day. Then it can use the predefined inspection district boundaries to assign the inspections to the various inspectors.

A county voter registration office might select the number of households that fall within each voting precinct. This number will help it determine how many poll workers each site might need.

School districts might map attendance boundaries for each of the elementary, junior high, and senior high schools. Then these boundaries might be used to determine how many students will attend each school, broken down by grade level, and the appropriate number of textbooks can be ordered.

5

Finding what's nearby

Not every important geographic relationship is based on overlapping or adjacent features. Many relationships are based on the idea that they are nearby or within a distance of each other. Near, however, is subjective; what is near to some may not be near to others. A near value then is associated with the type of data and the situation in which the data exists.

Tutorial 5-1

Selecting features nearby

One of the most powerful types of spatial analysis is finding features within a specified distance of other features. This measurement can be a straight-line distance, or distance along a network or path. Selected features can be displayed for visual analysis, or put into separate feature classes for use in other analyses.

Learning objectives

- *Select features*
- *Analyze distance relationships between features*

Preparation

- *Read pages 115–28 in* The Esri Guide to GIS Analysis, *volume 1.*

Introduction

In chapter 4, you used features to select other features. In this tutorial, you will use features and a distance value to find other features within that distance.

You will also discover that if all you need is a list or count of the nearby features, the process does not have to be complicated.

ArcMap has a selection tool called Select By Location, which you used in the previous chapter. A straight-line search distance can be added to the parameters of this tool to give it the added power to select things nearby.

This selection tool has two downsides. The distance is "as the crow flies," so it does not take into account street accessibility, terrain, or physical obstacles. Imagine if the analysis shows some great property within your search distance, but it is on the other side of a lake that will take five times the travel distance to get there.

The other drawback to the selection process is that the selection boundary is not drawn as a feature. The selected features are shown, but the boundary is not.

If the data does not have constraints such as those mentioned, and the selection boundary does not need to be shown, this process can be a quick and powerful analysis tool.

Scenario   The Tarrant County Health Department wants to come to Oleander to do some mosquito control work. Workers will walk the creeks, looking for standing water, debris that may be clogging the creek, and harborages for mosquito larvae. Before the workers arrive, however, the department wants to mail notices to all the property owners within 50 feet of the creek and let them know when the work will be done.

Data   The map document includes two layers. The first is the parcel dataset for the City of Oleander. Use the fields Addno, Prefix, StName, Suffix, and SufDir to get the street address for each selected parcel.

            The other dataset is the creek data from the North Central Texas Council of Governments, which maintains a lot of regional data.

Create a visual buffer

1   In ArcMap, open Tutorial 5-1.mxd.

5-1
5-2
5-3
5-4
5-5
5-6
5-7
5-8
5-9

        The map shows the first area that the county wants to clean up. The dashed purple line is the creek. You will use it as the selected feature and find parcels that are near it.

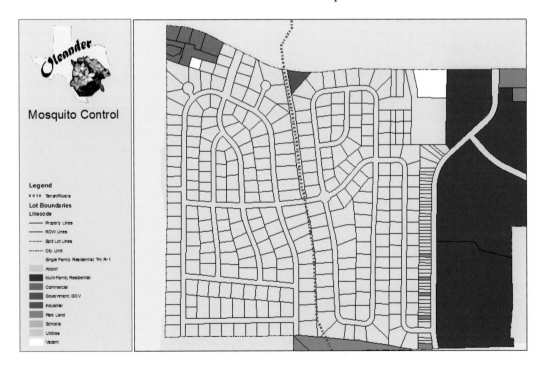

        There is a tool that will give a great visual of the target features. Try that first and see how it works. First, you will customize the standard ArcMap toolbar to include this tool.

2 On the main menu, click Customize > Customize Mode.

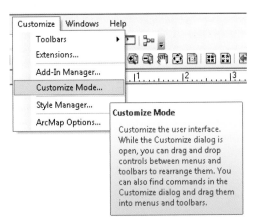

3 Go to the Commands tab, scroll through the Categories, and click Selection. In the right window, click the Buffer Selection tool  and drag it to the Tools toolbar. When you have it on the toolbar, release the mouse button. Close the Customize dialog box.

Note: when dragging a tool to a toolbar, a black bar will be displayed to show where the tool will be placed when released.

When you have time, scroll through the categories of tools in the Customize dialog box. These tools are not on the default toolbars but add great functionality to the program. Some of the commands can be found on menus. So if you use the commands frequently, it is helpful to add a button on the Tools toolbar. A search box on the Commands tab can help you find tools, or you can browse by category.

4 Click the new Buffer Selection tool. Select the TarrantRivers layer and set the distance to 50 feet. Then click OK.

5 Using the Select Features tool, click the purple creek feature to select it.

The Buffer Selection tool draws a 50-foot buffer around the selected features. Look closely at the creek and you will see the buffer.

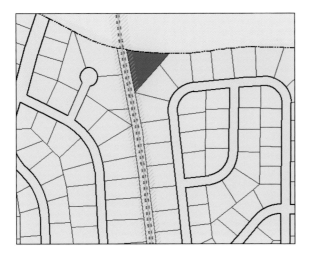

This buffer is great for visual analysis. When you print the document, the selection buffer will print. This will allow you to clearly see which parcels will be notified.

Unfortunately, it is a visual reference tool only. No features in the parcels dataset were selected, so there is no way to make a list. This tool is better suited for other purposes, so you will try another method.

6 Open the Buffer Selection tool again, select TarrantRivers, and set the buffer distance to zero. Click OK to close the dialog box.

In tutorial 4-2, you used the Select By Location tool. One parameter of this tool that you did not use is the ability to select within a distance of the selected features. The creek feature may still be selected from the last procedure. But if not, select it again.

Select by location using a distance buffer

1 On the main menu, click Selection > Select by Location to open the dialog box.

It is helpful to write the objective of the search so that you can understand what parameters to use in the selection dialog box:

I want to select features from the target layer Land Use that are within 50 feet of the features in the source layer TarrantRivers.

2 Fill out the selection dialog box as shown in the graphic. For Spatial selection method, select "intersect the source layer feature," and make sure that both the "Use selected features" and "Apply a search distance" boxes are checked. Set the distance to 50 feet. When your dialog box matches the graphic, click OK to make the selection.

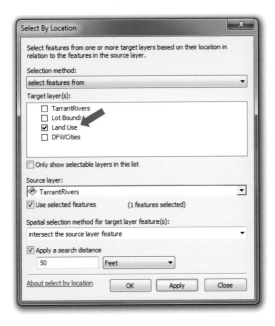

You now have a selection, but the cyan outlines will not look good on the printed map. You will change the selection color to red, but only for the Land Use layer.

Notice that even though you used a buffer distance in the selection, no buffer ring is drawn on the map.

Set the selection symbol and list the results

1 Right-click the Land Use layer and open the properties. Go to the Selection tab. Select the "with this symbol" option. Then click the cyan box to open the Symbol Selector dialog box. In the Symbol Selector, set the outline color to Flame Red and the outline width to 4. Click OK and then OK again to close the properties.

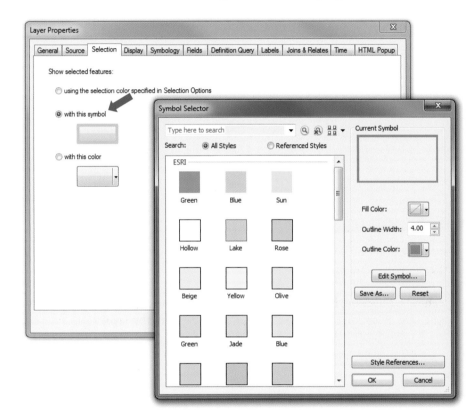

5-1

5-2

5-3

5-4

5-5

5-6

5-7

5-8

5-9

The map now shows clearly which parcels fall within the selection zone of the creek. Because you want only a list of addresses, you do not need these features to be in a separate feature class. You need only the list from the attribute table.

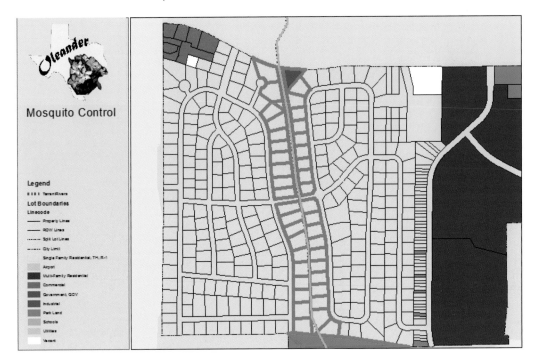

2 Right-click the Land Use layer and click Open Attribute Table. Click the Show selected records button 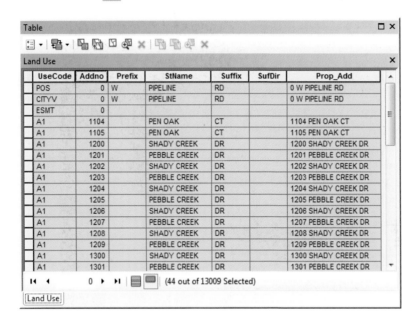 at the bottom of the table to view only the selected records.

The list can be exported to a DBF table and used to create mailing labels.

3 If you are working on this tutorial in a classroom environment, print the list for the completed project (Table Options > Print). When you are finished, close the attribute table, and click Selection > Clear Selected Features. Notice what happens to the map.

4 Save your map document as **Tutorial 5-1.mxd** to the \GIST2\MyExercises folder. If you are not continuing to the exercise, exit ArcMap.

5-1
5-2
5-3
5-4
5-5
5-6
5-7
5-8
5-9

Exercise 5-1

The tutorial showed how to use a selected feature, and the Select By Location tool, to select land-use parcels within 50 feet of the selected creek.

In this exercise, you will repeat the process for one of the other creeks that the county wants to inspect.

- Continue using the map document you created in this tutorial, or open Tutorial 5-1.mxd from the \GIST2\Maps folder.
- Go to the bookmark Creek Area 2.
- Select the creek in that region.
- Use the Select By Location tool to select all the land-use parcels within 50 feet.
- Print the resulting list of addresses.
- Change elements such as the titles, colors, and legend to make a visually pleasing map.
- Save the results as **Exercise 5-1.mxd** to the \GIST2\MyExercises folder.

What to turn in

If you are working in a classroom setting with an instructor, you may be required to submit the maps you created in tutorial 5-1.

Turn in a printed map or screen capture of the following:

Tutorial 5-1.mxd and a printed list of the selected addresses

Exercise 5-1.mxd and a printed list of the selected addresses

Tutorial 5-1 review

The selection processes in this tutorial used a straight-line distance from the selected features to select other features. The software temporarily creates a buffer zone, and features that touch that zone are selected. The buffer is never seen, and cannot be drawn on a map, but exists only for the moment to make the selection.

The selected set can be used to get lists or counts of features within the distance. But because no hard boundaries are drawn, it limits what can be done with the selection. If the desired result is to measure areas or calculate densities, use other more advanced tools to do it.

Study question

1. When is it appropriate to perform summary operations against the selected data?

5-1
5-2
5-3
5-4
5-5
5-6
5-7
5-8
5-9

Other real-world examples

The public works department might use a straight-line distance selection to get a list of property owners along a street that is being scheduled for repairs.

The health department might identify a hazardous site such as an abandoned lead smelter. It can select houses within a certain distance of the site to alert residents of upcoming cleanup efforts.

The economic development department might select census tracts around a business center to compile demographic data about the area.

Tutorial 5-2

Creating buffer features

A straight-line distance buffer can be created around selected features. This new feature can be symbolized on the map and used in overlay analysis to clip other features.

Learning objectives

- Create buffers
- Use buffers to select features

Preparation

- Review pages 115–28 in The Esri Guide to GIS Analysis, volume 1.

Introduction

Buffers can be created on a more permanent basis than the selections you did in tutorial 5-1. They can be saved as their own feature class, allowing you to do the same type of inside-outside analysis as before. The difference now is that the region you are using was created by applying a distance to a selected set of features.

Again, you have the choice of three overlay analysis tools to use with the buffers: Identity, Intersect, and Union. Review these tools in ArcGIS for Desktop Help to determine their functionality.

Once you do the overlay function, you can use the resulting feature class for calculations such as sum of area. This function also creates a feature on the map that you can symbolize to show the area of study.

Scenario

Every time a zoning change is proposed, the property owners within 200 feet of the subject tract must be notified by mail. They are given the date and time of the city council meeting at which the change will be discussed. You have been told of a certain tract of land in which a zoning change is being proposed, and the city secretary is requesting a list of adjacent property addresses.

The process is simple. You will select the subject tract, buffer it 200 feet, and select the parcels that intersect the buffer. You can do this task using only a selection, but the map must have a graphic that shows the notification boundary.

Data

All you need for this tutorial is the parcel dataset for the City of Oleander. Use the fields Addno, Prefix, StName, Suffix, and SufDir to get the street address for each selected parcel.

Create a buffer

1   In ArcMap, open Tutorial 5-2.mxd.

The purple-hatched parcel is the one that is subject to a zoning change. The property owner wants to change from C-2 (Community Business District) to C-2(A) (Community Business District with Alcohol Sales) for a new convenience store.

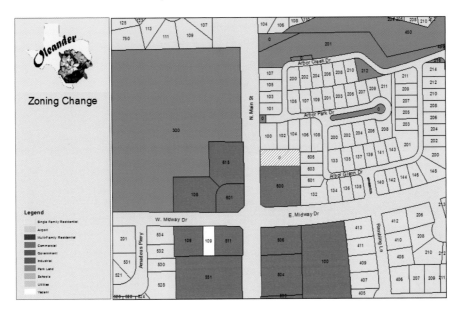

The first thing you will do is create a 200-foot buffer around the subject tract. The tract is already selected, and the selection color is the purple hatch that you see.

2   Click the Search tab and enter **Buffer**. Click Buffer (Analysis) to open the buffer dialog box.

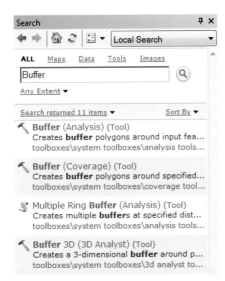

5-1
5-2
5-3
5-4
5-5
5-6
5-7
5-8
5-9

3 In the Buffer dialog box, click the down arrow for Input Features and select the Land Use layer.

4 Name the output feature **ZoningCase1** and save to MyData.gdb in the \GIST2 \MyExercises folder.

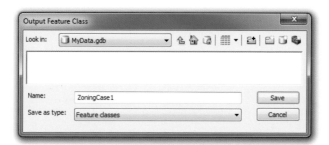

5 Set the linear distance to 200 and the units to Feet. When your dialog box matches the graphic, click OK.

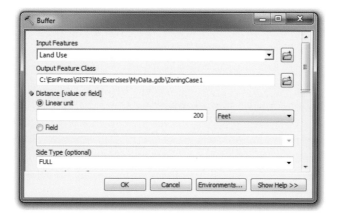

A buffer feature is created around the selected parcel. Notice that it is not a circle! The Buffer tool buffers the edge of the parcel polygon, so the resulting feature takes on the shape of the original feature. This is the desired result, but you can make it prettier for your map.

Update the symbology

1 Right-click the ZoningCase1 layer and open the properties. Go to the Symbology tab.

2 Click the Symbol box to open the Symbol Selector and select Lilac. Click OK.

3 Go to the Display tab. Set the Transparent value to **60**%. Click OK to close the Properties dialog box.

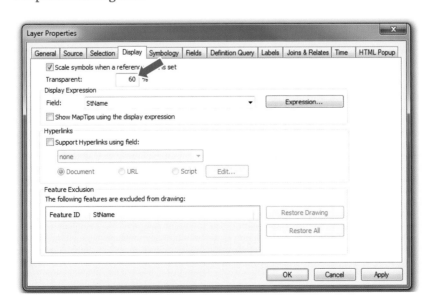

Now the map looks better. It clearly delineates the subject tract, and the 200-foot boundary.

4　Print or save your map document as **Tutorial 5-2a.mxd** to the \GIST2\MyExercises folder before you move on to the selection step, since the selection will turn a lot more parcels purple.

The next step is to perform the selection process. Once again, you will use the Select By Location tool and build a selection sentence. You do not need to specifically select the buffer feature because it is the only feature in that feature class. You also do not need to use a distance with the selection because the buffer already represents the 200-foot buffer that you need.

The selection sentence is as follows:

I want to select the features from the Land Use layer that intersect the features in the ZoningCase1 layer.

View the attributes of selected features

1 On the main menu, click Selection > Select By Location. Determine on your own what the selection parameters will be; then configure the Select By Location tool. When you have confirmed that your dialog box matches the graphic, click OK.

The selected features turn purple because of the changes you made to the selection color in the layer properties. All you need is the list from the attribute table.

2 Right-click the Land Use layer and click Open Attribute Table. Click the Show selected records button to view only the selected features.

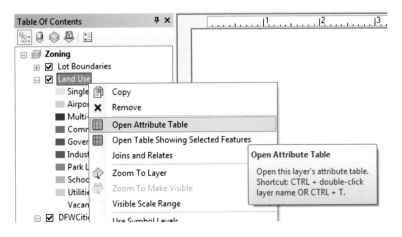

Customizing a menu

Notice the extra tool on the menu in the previous graphic in step 2: Open Table Showing Selected Features.

This tool saves the step of clicking the Selected button after opening a table. You can add the tool (if you do not already have it) in the same manner as the Buffer Selection tool in tutorial 5-1.

- On the main menu, click Customize > Customize Mode.
- Click the Commands tab on the Customize dialog box and scroll to the Layer category.
- On the Toolbars tab, select the Context Menus check box to open the Context Menu toolbar. Click the down arrow next to the Context Menus button and select Feature Layer Context Menu.
- Find Open Table Showing Selected Features in the Commands list and drag it to the Feature Layer context menu.

Close the Customize Mode dialog box. Right-click a layer to test the new command.

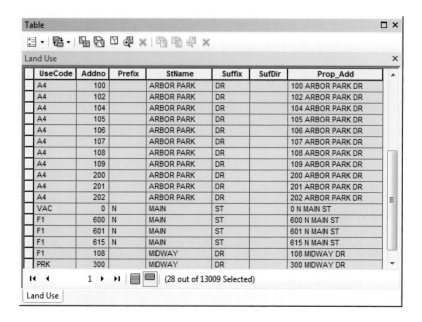

The resulting table can now be used to create mailing labels, printed to show a list of properties, or exported to a table and saved for future use.

Exporting a table

The results of this analysis can be useful in a spreadsheet, or used to perform a mail merge using word-processing software. To export only the tabular data of a layer, open the layer's attribute table and click the Table Options button . There you will find the Export tool. The data can be exported as a personal or file geodatabase table, as a dBase file, or as a delimited text file. Notice that the Export drop-down menu will allow you to export all the records, or only the currently selected features.

3   Print or save your map document as **Tutorial 5-2b.mxd** to the \GIST2\MyExercises folder. If you are not continuing to the exercise, exit ArcMap.

5-1
5-2
5-3
5-4
5-5
5-6
5-7
5-8
5-9

Exercise 5-2

The tutorial showed how to create a buffer feature from the selected features. A distance is specified, and a new feature is created by extending the boundary of the selected feature by the given distance.

In this exercise, you will repeat the process using a different parcel. The property owner who requested the zoning change in the tutorial has selected a second lot for another location. The request is to also change from C-2 (Community Business District) to C-2(A) (Community Business District with Alcohol Sales).

- Continue using the map document you created in this tutorial, or open Tutorial 5-2.mxd from the \GIST2\Maps folder.
- Go to the bookmark Zoning Case 2.
- Select the vacant lot on the southwest corner of North Main Street and Bear Creek Drive.
- Create a 200-foot buffer feature around the subject tract.
- Select the parcels that will be notified of the impending zoning change.
- Save the results as **Exercise 5-2.mxd** to the \GIST2\MyExercises folder.

What to turn in

If you are working in a classroom setting with an instructor, you may be required to submit the maps you created in tutorial 5-2.

Turn in a printed map or screen capture of the following:

Tutorial 5-2a.mxd, Tutorial 5-2b.mxd, and a printout of the selected parcel addresses

Exercise 5-2.mxd and a printout of the selected parcel addresses

Tutorial 5-2 review

Unlike a straight-line distance selection, this time you created a feature to use for the selections. The buffer you created is the result of a straight-line measurement from the selected feature, whether it is a point, a line, or a polygon.

With the buffer as a separate feature, things can be done with the buffer that cannot be done using the straight-line selection process. Most important, the buffer can be symbolized on the map. Also, because it is a feature, its area can be measured.

Another important difference can be seen if the buffer is done on multiple features. The buffers can each produce a unique selection set, whereas the straight-line selection produces only a single set of features as a result.

Study questions

1. Will the selection with a buffer always yield the same results as the straight-line selection?

2. Name three characteristics of a buffer feature that a straight-line selection does not have.

Other real-world examples

The police department buffers schools, playgrounds, and arcades to produce zones of strict drug enforcement. Each of these types of facilities is given a different buffer size, which cannot be done using a straight-line selection. The buffers are also depicted on a map to show field officers the boundaries of the zones.

The local hospital buffers its facilities 500 feet to produce a quiet zone. A printed graphic of the zones is distributed to local citizens. The police department may also use it to help enforce the restrictions.

5-1
5-2
5-3
5-4
5-5
5-6
5-7
5-8
5-9

Tutorial 5-3

Clipping features

Besides providing a symbolized graphic on the map, a buffer layer can also be used with the overlay analysis tools. The buffer can be used to cut features out of other layers, allowing you to determine the exact area of the features inside the buffer.

Learning objectives

- *Select features*
- *Create buffers*
- *Intersect features*
- *Dissolve features*
- *Analyze overlays*

Preparation

- *Review pages 115–28 in* The Esri Guide to GIS Analysis, *volume 1.*

Introduction

The feature created with the Buffer tool is only that, a feature. As with any other polygon, it can be used with the overlay analysis tools. Features can be clipped using the buffer ring like a cookie cutter. The resulting features are clipped at the boundary and have a new area.

Using the selection tools, you can only get lists and counts of features. Once the features are clipped, you can do summaries on the area, and with density values you can calculate the proportion or percentage of a value within the buffer.

As you saw in tutorial 4-2, there are many tools used in overlay analysis to cut the sets of input features according to different rules. In that exercise, the Union tool was used to bring two datasets together because it was important to preserve all the features and all the attributes from both datasets. This tutorial uses Intersect, which preserves only the common areas of the features but saves all the attributes from both datasets. In contrast, the Clip tool can be used to get the common areas, but it does not preserve all the attributes from both datasets. Before you continue, review the tools in the Overlay toolbox to gain a good understanding of the characteristics of these tools.

Scenario   A homeowners group next to the subject tract has challenged the zoning notification process from tutorial 5-2. All the owners of the residential lots signed the petition to block the zoning change. The city council must still vote and decide the issue. However, if 60 percent of the area inside your notification buffer is residential, the city council vote will have to be a supermajority (7 to 2) for the ordinance to pass.

You will provide the city attorney an exhibit that shows the percentage of each land-use type inside the notification zone. He will advise the city council of the vote count that is needed to approve the requested zoning change.

Data   All you need for this tutorial is the parcel dataset for the City of Oleander. You will use the fields UseCode and Shape_Area to get the data you need.

Dissolve a layer

1   In ArcMap, open Tutorial 5-3.mxd.

Notice the buffer feature you created in tutorial 5-2. You will use the buffer feature as a cookie cutter to clip out the parcels that fall within it.

5-1
5-2
5-3
5-4
5-5
5-6
5-7
5-8
5-9

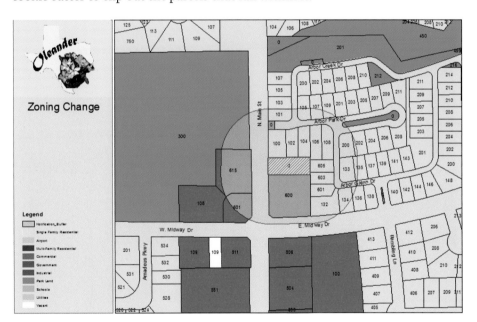

This time you cannot use the Select Features tool, because parcels that are partially inside the buffer will be selected in their entirety. Instead, you will use the Intersect tool to cut out the desired parcels. The resulting parcel segments can be used to get an exact area inside the buffer.

When the Intersect tool is run, it will act upon only the selected features—unless no features are selected, in which case it will act upon the entire dataset. This rule is true of all the geoprocessing tools, so knowing how to predict this action can be useful when designing processes. Imagine having a countywide dataset from which you want to do an intersect in only a certain region. If you select that region first, the unselected features will not be used in the intersect process, thus saving a lot of processing time. In this case, it is okay to have the Intersect tool use the entire Parcels dataset. If you want to check for selected features, go to the Selection tab at the bottom of the table of contents and see which layers have selected features. In this analysis, it is best to start with no selected features.

2 On the Tools toolbar, click the Clear Selected Features button if necessary to ensure that no features are selected.

3 Open the Search tab and type **Intersect**. Click the tool in the Results window to run it.

The Intersect tool outputs a new feature class that contains the areas that are common to both input layers.

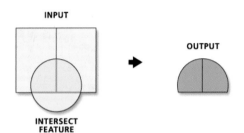

The Intersect tool computes a geometric intersection of the Input Features. Features or portions of features that overlap in all layers and/ or feature classes will be written to the Output Feature Class.

4 In the Input Features drop-down list, select Land Use to add it to the list of layers.

5 Repeat the process to add the Notification_Buffer layer to the list.

6 Save the output feature class as **ZoningCase1LandUse** in MyData.gdb in the \GIST2\MyExercises folder. When your dialog box matches the graphic, click OK to run the Intersect process.

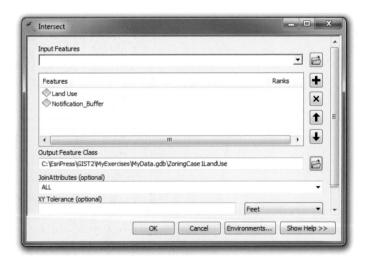

A new layer is created and added to the map document. This layer is the clipped features that are common to both input layers.

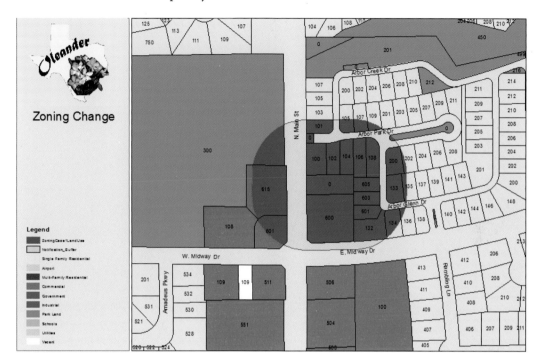

You will get the percentage of each land-use type in this new coverage. To get the percentage, you will dissolve the layer by the UseCode field, creating a new feature class with one polygon for each unique land-use code. Each polygon will have an area calculated for it, and you can use that area to get the percentage.

7 Use the Search window to find and open the Dissolve (Data Management) tool.

8 In the Dissolve dialog box, set Input Features to ZoningCase1LandUse.

9 Name the output file **LandUseDissolve** in MyData.gdb in the \GIST2\MyExercises folder.

5-1
5-2
5-3
5-4
5-5
5-6
5-7
5-8
5-9

10 Set the Dissolve field to UseCode by selecting the check box next to it in the list. Note that the UseCode field was duplicated in the intersect process because it existed in both input feature classes. When your dialog box matches the graphic, click OK to start the dissolve process.

Analyze the dissolved layer's data

The dissolve process creates a new feature class, but in reality you only need the data it contains. The attribute table of this new layer contains the area of each unique occurrence of UseCode in the original file. You will look at the table, and then add a field to calculate the percentage of the total. Remember that if the residential category exceeds 60 percent, the city attorney will require a supermajority vote at the city council meeting.

1   Right-click the LandUseDissolve layer and click Open Attribute Table. Note the
UseCode field values, and the ArcMap-generated Shape_Area field. Right-click the
Shape_Area field and click Statistics. Make a note of the Sum value. Yours may be
slightly different from the value shown in the graphic. Close the Statistics window.

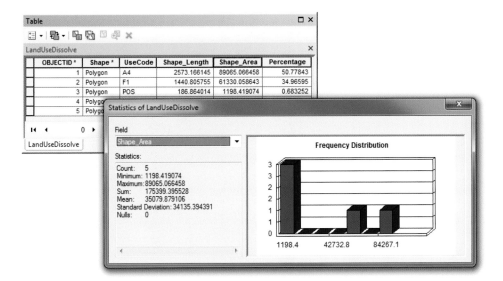

2   Click the Table Options button at the top of the attribute table and click Add Field. In
the Add Field dialog box, type **Percentage** for Name. Click the Type down arrow and
select Float. Click OK to create the field.

3   In the attribute table, right-click the new Percentage field and click Field Calculator.
Click Yes if you receive a message about calculating outside an edit session.

5-1
5-2
5-3
5-4
5-5
5-6
5-7
5-8
5-9

4 In the Field Calculator dialog box, double-click the Shape_Area field to add it to the calculation. Type the rest of the calculation as **/ 175399.395528 * 100**. Click OK.

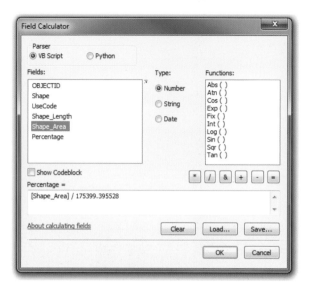

5 Click OK to calculate the new field values.

Change a table's appearance

You can see from the table that the percentage of residential property (A4) does not exceed 60 percent. Your value may differ from the value in the graphic. This data can be added to the map layout, but first you will clean it up and set some symbology so that it will look nice.

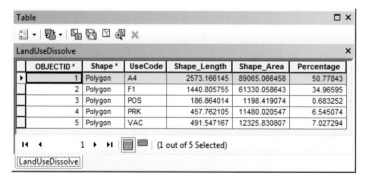

1 At the top of the table, click the Table Options button and click Appearance.

2 Set the table font to Arial Black and the size to 10, and then click OK to finish.

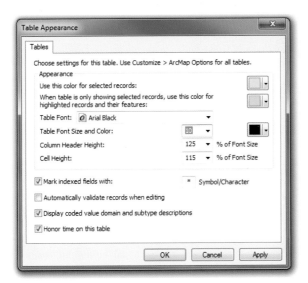

This makes the font easy to read on the final map, but you do not want all the extra fields to be visible. You will turn some of them off, and change their names to aliases that make more sense. You will also format the numbers for easier readability.

3 Move the table out of the main map window or dock it to the bottom of the display. Open the properties of the LandUseDissolve layer and go to the Fields tab. Clear all the fields, except UseCode and Percentage.

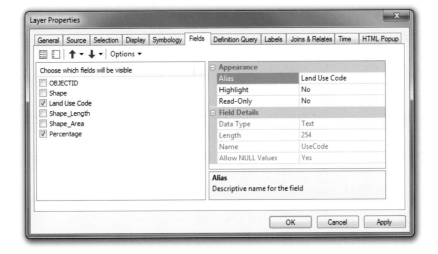

5-1
5-2
5-3
5-4
5-5
5-6
5-7
5-8
5-9

4 Highlight the UseCode field. In the Appearance pane, change the alias to **Land Use Code**. Then highlight the Percentage field and make its alias **% of Total Area**.

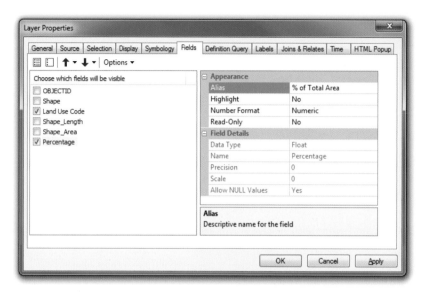

5 Highlight the Number Format line and click the Number Format button. In the resulting window, set the number of decimal places to 1.

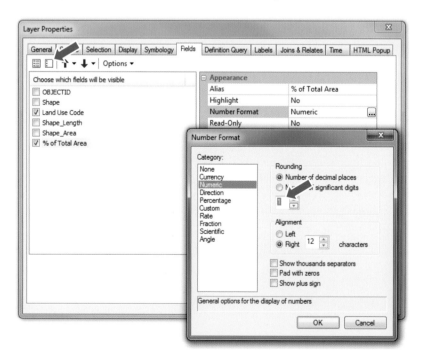

6 Click OK and then OK again. Resize the field widths and table box so that only the data is visible.

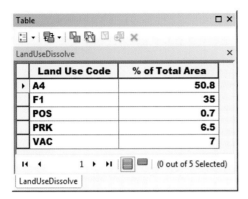

7 Click the Table Options button and click Add Table to Layout. Then close the table.

8 The table is dropped into the middle of the map. Drag it to the left and place it on the title bar. Make sure that the only visible layers are Land Use, Notification_Buffer, DFWCities, and Lot Boundaries.

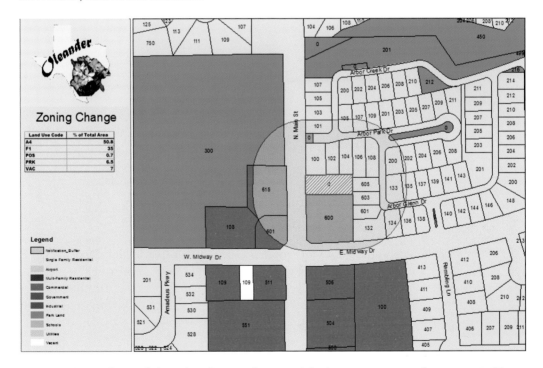

Much of the work you did on this data, and some of the layers you created, are not visible on the final map. They are important steps in calculating the desired result: the percentage of land use inside the buffer. However, they do not add much graphically to the map layout.

The percentage of residential land use in the area did not require the city council to achieve a supermajority vote to either pass or deny this request. Oddly enough, the city council unanimously voted down the request because of what else was nearby, as demonstrated in your exhibit: two city parks. The land remains vacant.

9 Save your map document as **Tutorial 5-3.mxd** to the \GIST2\MyExercises folder. If you are not continuing to the exercise, exit ArcMap.

Exercise 5-3

The tutorial showed how to clip a layer using a buffer as the boundary. The resulting table is used for a generalization operation to extract the necessary data.

In this exercise, you will repeat the process using a different parcel. With the original project denied, the applicant is putting all efforts into getting a second-choice site approved. The homeowners in that area have also filed a petition to force a supermajority vote, and the city attorney has asked for the same percentage breakdown of land-use types.

- Continue using the map document you created in this tutorial, or open Tutorial 5-3.mxd from the \GIST2\Maps folder.
- Go to the bookmark Zoning Case 2.
- You will work with the vacant lot on the southwest corner of North Main Street and Bear Creek Drive.
- Use the buffer created in exercise 5-2, along with the Intersect and Dissolve tools, to derive the necessary information.
- Create a summary table to include in the layout.
- Save the results as **Exercise 5-3.mxd** to the \GIST2\MyExercises folder.

What to turn in

If you are working in a classroom setting with an instructor, you may be required to submit the maps you created in tutorial 5-3.

Turn in a printed map or screen capture of the following:
> **Tutorial 5-3.mxd**
> **Exercise 5-3.mxd**

5-1
5-2
5-3
5-4
5-5
5-6
5-7
5-8
5-9

Tutorial 5-3 review

Buffers used for selections can only provide lists and counts, but with the overlay tools it is possible to start calculating areas of intersection. The same type of buffer is created as the selections, but now the overlay tools can be used to clip out only the portions of features that fall within the buffer. Once these features are placed into a new feature class, they can be manipulated separately from the original features.

It is important to note that any summary values that might be true of the whole piece are no longer representative of the area. If a census block with a value of 200 people is split in two, each half will still retain the value of 200. Summary statistics must be calculated as densities before the overlay operation to retain their relationship to the new area.

Study questions

1. What is the difference between Intersect and Identity?

2. When is Union the best choice?

Other real-world examples

The health department might buffer water wells to show a spill hazard area and use the buffer to clip out the soil polygons. The clipped features can be used to get a total number of different soil types in the hazard area to evaluate soil permeability.

The regional water authority may buffer a lakeshore by the distance of an access easement and use the buffer to intersect property records. The resulting clipped features can be used to calculate the percentage of the easement that each property owner owns.

Tutorial 5-4

Buffering values

Buffering can be done with a single straight-line distance for all features. Or, with an optional setting, each feature can be buffered by a separate distance value. These values are stored in an attribute field and can be referenced in the buffer command.

Learning objectives

- *Work with buffers*
- *Work with attribute values*
- *Perform distance analysis*

Preparation

- *Review page 124 in* The Esri Guide to GIS Analysis, *volume 1.*

5-1
5-2
5-3
5-4
5-5
5-6
5-7
5-8
5-9

Introduction

The single-distance buffer allows you to define one distance for all the features. The buffer command includes an option to derive the buffer distance from a field in the attribute table. So every feature can have a unique buffer distance.

When the buffers are drawn, the option of merging the output features becomes critical. Not merging the buffers will create a separate polygon for each feature in the input file. The result can be many polygons. The option to merge them will dissolve the boundaries where the buffers overlap and make fewer polygons with larger areas.

The decision to merge may depend on the ultimate use of the buffers. For visual analysis, the dissolve method looks best. But if a one-to-one relationship between the buffers and the origin features must be preserved, the dissolve function should not be used.

Imagine a stream network with continuous and intermittent streams, each needing a buffer of a different distance. The required distance can be saved in a field and used in the Buffer tool. At the point in which two streams come together, the buffers can be set to merge, with one buffer being derived from several stream segments. If it is important to preserve the names of the streams in the output buffer features, the dissolve will not be used and a one-to-one relationship can be maintained.

Scenario   The city council has received a report on the effects of noise pollution caused by cars driving on city streets. To get an idea of how much of the city might be affected by street noise, the council has asked you to prepare an exhibit map. A consultant has sent the noise coefficients for each street in the city, and they have been put into a field in the attribute table. You will buffer the street centerlines to visually show the area affected by vehicle noise.

Data   The street centerline data contains a field named CO_CODE that represents the distance that noise of a high-decibel level travels from each street.

The parcel layer is included for background interest.

Search for a tool

1   In ArcMap, open Tutorial 5-4.mxd.

Here is a plain-looking map. The street centerline file is included in the table of contents, but it is not visible. You do not need to see it; you will only use it in a buffer analysis.

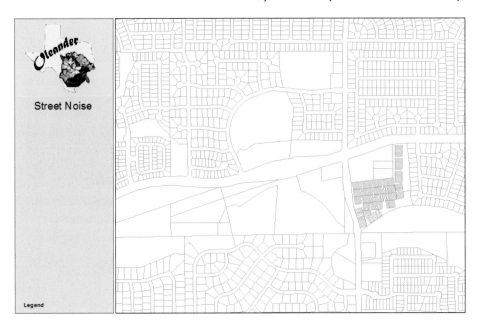

2   Open the Search window and type **buffer**. Click the Buffer (Analysis) tool in the Results window.

Set buffer parameters

1   In the Buffer dialog box, set Input Features to StreetCenterlines.

2   Set Output Feature Class to \GIST2\MyExercises\MyData.gdb, and name the feature class **NoiseBuffers**.

3 Under Distance, click the Field option. In the drop-down list, set the field to CO_CODE.

4 Set Dissolve Type to ALL. When your dialog box matches the graphic, click OK.

As expected, you get a wider buffer on streets that create more noise and a smaller buffer on streets that generate less noise. The city council will now be able to visualize the table of values that the consultant delivered.

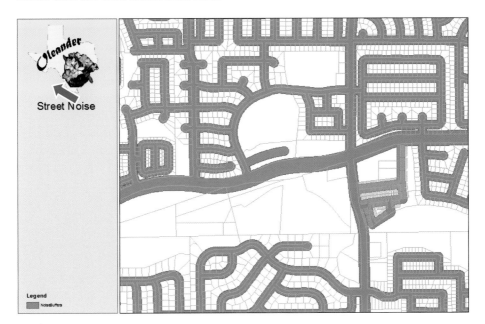

5-1
5-2
5-3
5-4
5-5
5-6
5-7
5-8
5-9

5 Save your map document as **Tutorial 5-4.mxd** to the \GIST2\MyExercises folder. If you are not continuing to the exercise, exit ArcMap.

Exercise 5-4

The tutorial showed how to create a multidistance buffer using a value from a field in the attribute table.

The city council reviewed your map, but it wants to see the relationship between speed limit and the measured noise values. In this exercise, you will create a set of multisize buffers using the speed limit field. Overlay the results on top of the noise buffers for a visual comparison.

- Continue using the map document you created in this tutorial, or open Tutorial 5-4.mxd from the \GIST2\Maps folder.
- Use the Buffer tool and the field SpeedLimit to create the buffers.
- Symbolize and arrange the layers to make the best map possible.
- Save the results as **Exercise 5-4.mxd** to the \GIST2\MyExercises folder.

What to turn in

If you are working in a classroom setting with an instructor, you may be required to submit the maps you created in tutorial 5-4.

Turn in a printed map or screen capture of the following:

Tutorial 5-4.mxd

Exercise 5-4.mxd

Tutorial 5-4 review

Although this process also creates a straight-line buffer, it is unique in its ability to vary the size of the buffer. Some data preparation must be done to create a field and populate it with a buffer distance.

The field chosen for the buffers must be a quantity that is measured in the current map units. Quantities that do not represent a measurable distance, such as traffic counts or rankings of road conditions, are not suitable. Codes that represent a quality of the road are also not suitable because they do not represent a distance. It is also important to note that the field must be represented in map units, since the Buffer tool does not allow unit conversions. So if the map units are meters, be careful not to use values measured in feet for the buffers.

The different-size buffers can also be created as single pieces as opposed to merging them all together. Thus, there is a greater number of features, but more detailed selections and overlays can be done using a subset of the buffer features. For instance, all the buffers representing the loudest noise can be totaled to show a percentage of the overall impact.

Study questions

1. Can buffers be generated around a text field, such as a code?

2. How can you solve the problem of mismatched buffer value units and map units?

Other real-world examples

The public works department might buffer each streetlight in the city with a different value based on its wattage. Using buffers can provide a visual map of areas that are well lit, and those that are not.

The local electrical utility might buffer its power lines by different values depending on the lines' voltage. The buffers may be used to determine areas required for clearance easements and the value of the included property.

5-1
5-2
5-3
5-4
5-5
5-6
5-7
5-8
5-9

Tutorial 5-5

Using multiple buffer zones

The straight-line buffer command can be repeated several times to analyze several distances, but there is a tool that allows multiple distances to be entered for the buffering process.

Learning objectives

- *Create multiple-ring buffers*
- *Perform distance analysis*

Preparation

- *Review page 126 in* The Esri Guide to GIS Analysis, *volume 1.*

Introduction

The Multiple Ring Buffer tool allows the user to input several distances at one time to create multiple buffers. On the surface, it looks as if the regular Buffer tool is running multiple times, but the multiple-ring buffer command has some key features that distinguish it.

When the multiple-ring buffers are created, a dissolve option is presented. Not dissolving will make a full circle feature for each distance provided. The features will overlap. If these individual buffers are used to do an overlay analysis, selected features can reside in several of the buffer rings created. In fact, features at the center of the buffered area will be inside all the buffer rings.

If the output multiple-ring buffers are dissolved, the features resemble a doughnut with the interiors clipped out where they overlap the other rings. As a result, the overlay analysis will place the features in only one ring. They will be farther than one distance, but nearer than the next distance.

The other important difference is that the output buffer rings contain a field that shows the buffer distance for each ring.

It is also important to note that the distance values can be in any sequence. They do not have to be at regular intervals.

Scenario   The fire chief wants you to find the number of houses within certain distances of each fire station. The chief figures that a fire truck averages about 30 mph on a fire run, and wants you to create rings that represent one minute, two minutes', and three minutes' drive time from each fire station.

You will create these rings and do an overlay analysis on the building footprints.

Data   The chief provided you with the fire station locations. Each has a station number with this code:
- 551 = Station 1
- 552 = Station 2
- 553 = Station 3

Building footprints were compiled from aerial photos taken last year, with a polygon that represents each building. Usage codes, called UseCode, represent how each building is used:
- 1 = Single Family Residential
- 2 = Multi-Family Residential
- 3 = Commercial
- 4 = Industrial
- 5 = Government
- 6 = Utilities
- 7 = Schools
- 8 = Churches

The parcel data is added for visual interest, but it will not have an effect on the analysis.

5-1
5-2
5-3
5-4
5-5
5-6
5-7
5-8
5-9

Create a multiple-ring buffer

1 In ArcMap, open Tutorial 5-5.mxd.

Shown are the parcel lines, building footprints, and Station 1, symbolized by a blue box. You will select the point that represents the station and buffer it. By default, the Buffer tool will buffer only the selected features.

2 Using the Select Features tool 🔍 , select the point that represents Fire Station 1.

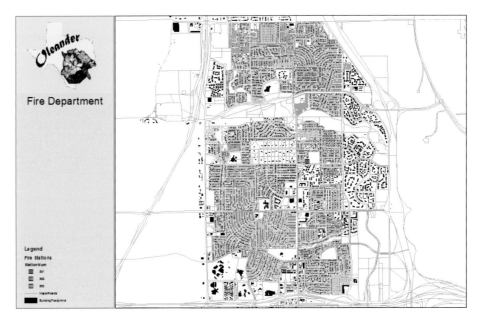

3 In the Search window, locate and open the Multiple Ring Buffer tool.

4 Set Input Features to Fire Stations.

5 Set Output Feature class to \GIST2\MyExercises\MyData.gdb and name the feature class **Station1Buffers**.

The fire chief gave you an average vehicle speed and the times of one, two, and three minutes. Doing the math gives you distances of 2,640 feet; 5,280 feet; and 7,920 feet, respectively.

6   Enter the distance **2640** and click the plus sign to add the distance to the list.

5-1
5-2
5-3
5-4
5-5
5-6
5-7
5-8
5-9

7   Type the distances **5280** and **7920**, clicking the plus sign after each one to add them to the list. Verify that the buffer distance is set to feet.

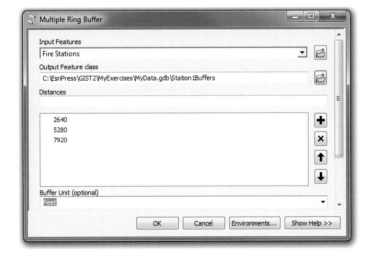

8 Scroll down to the bottom of the dialog box and verify that Dissolve Option is set to ALL. When your dialog box matches the graphic, click OK.

Select by location using buffer rings

The resulting buffer rings are added to the map document. The new layer must be at the bottom of the table of contents so that the other layers draw over the rings.

1 Drag the new buffer layer to the bottom of the table of contents.

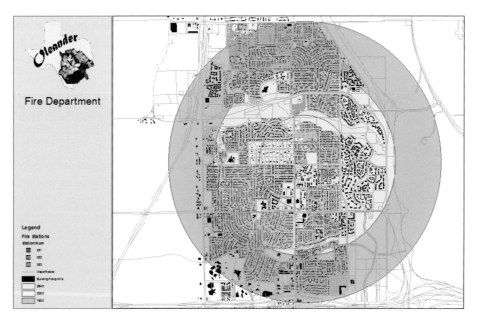

By using the Dissolve option, the rings will not overlap but be more of a doughnut shape. You will check that by selecting the outer ring.

2   Clear the selected features and make the new buffer layer selectable. Using the Select Features tool, select the outer ring of the buffers.

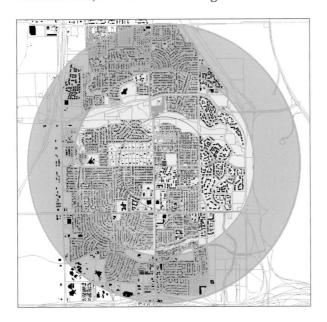

5-1
5-2
5-3
5-4
5-5
5-6
5-7
5-8
5-9

The significance of the doughnut ring is that you will be able to use it in your overlay processes to select features that are at least one distance away, but not farther than another distance. In this instance, you can select the housing units that are at least 5,280 feet away, but closer than 7,920 feet. You must be careful to set your selection method so that houses that straddle the boundary do not get selected in two rings.

3   On the main menu, click Selection > Clear Selected Features, or click the Clear Selected Features button on the Tools toolbar.

4   With the Select Features tool, select the innermost ring of the buffers.

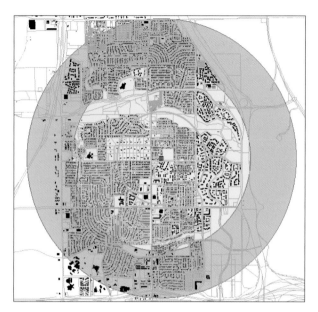

Once again, you are going to use one feature to select others using the Select By Location tool. You will want to select features from the building footprints that have some spatial relationship with the selected features of the Station1Buffers layer. But what selection method should be used? Look at the choices offered under "Spatial selection method for target layer feature(s)" and determine if any of these methods can be used to select each building only once for each of the buffer rings:

Intersect—buildings that straddle the boundary are selected in two rings. (No)

Are within a distance of—you are using the buffer features, not the selection distance, to select buildings. (No)

Contain—this is a reverse selection. The buildings must contain part of a ring. (No)

Completely contain—this is also a reverse selection. The buildings must contain an entire ring. (No)

Contain (Clementini)—the buildings must contain an entire ring without sharing the border. (No)

Are within—if any part of the building is inside one of the rings, it could be selected in two rings. (No)

Are completely within—again, think of the features that straddle the boundary that would be selected in two rings. (No)

Are within (Clementini)—if any part of the building is within the rings, unless it shares the complete boundary, it could be selected in two rings. (No)

Are identical to—the outline of the building must be identical to one of the rings. (No)

Touch the boundary of—only the buildings that straddle the boundary are selected. (No)

Share a line segment with—the building's line work must be coincident with the ring's arc. (No)

Are crossed by the outline of—only the buildings that touch the boundary are selected. (No)

Have their centroid in—because the centroid point can occur in only a single ring, each building is selected with only one ring; no double selection. (Yes)

So after looking at all the choices, you will use the spatial selection method "Have their centroid in." This method ensures that each building footprint is counted in only one ring.

5 On the main menu, click Selection > Select By Location to start the process. You have used this method before, so try completing the dialog box on your own. When it matches the graphic, click OK.

5-1
5-2
5-3
5-4
5-5
5-6
5-7
5-8
5-9

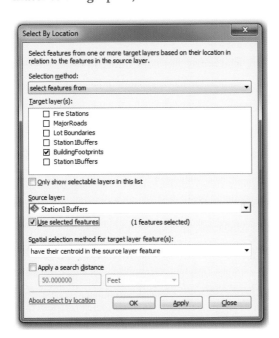

How many features were selected? How do you know? Where can you find out?

6   Go to the List by Selection button at the top of the table of contents. The number next to BuildingFootprints is the number of currently selected features. Make a note of the number.

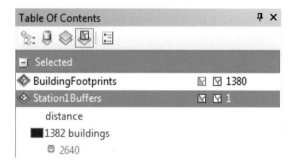

YOUR TURN

Perform the Select By Location process for the remaining two buffer rings. Make a note of the total number of structures for each distance—you will use them in step 7.

7   Change the symbology of the Station1Buffers layer to Quantities > Graduated colors with distance as the Value field. Choose a light-blue to dark-blue color ramp, and then flip the ramp so that the zone nearest to the fire station has the darkest shade. To flip the ramp, click the Symbol column heading and click Flip Symbols. Change the labels to reflect the number of structures in each buffer zone. When your dialog box matches the graphic, click OK. Finally, move the buffer layer below the BuildingFootprints layer.

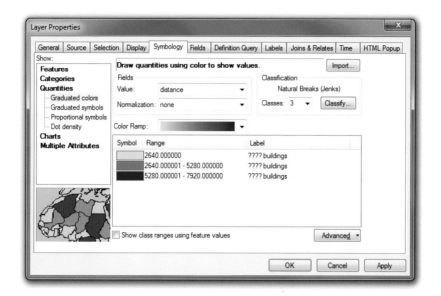

The blue shades were chosen to go along with the blue dot chosen for Station 1. Incorporating the selection results into the legend's color scheme keeps the map layout simple in form yet still conveys the information clearly.

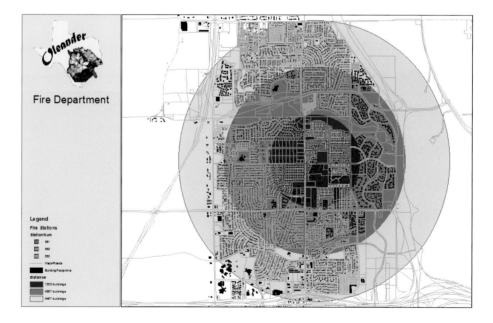

8 Save your map document as **Tutorial 5-5.mxd** to the \GIST2\MyExercises folder. If you are not continuing to the exercise, exit ArcMap.

Exercise 5-5

The tutorial showed how to create multiple search distances using the Multiple Ring Buffer tool. Each ring becomes a selection feature for a Select By Location process.

In this exercise, you will repeat the process for the other two stations. For both Station 2 and Station 3, create a map that displays the number of structures within each of the three response times. For each set of buffers, use a color scheme that complements the corresponding station.

- Continue using the map document you created in this tutorial, or open Tutorial 5-5.mxd from the \GIST2\Maps folder.
- Go to the bookmark Station 2.
- Create the multiple-ring buffers.
- Perform the selections for each ring.
- Set all the symbology for the layers, as well as any other title changes or notes, to make the map visually appealing.
- Repeat for Station 3.
- Save the results as **Exercise 5-5.mxd** to the \GIST2\MyExercises folder.

What to turn in

If you are working in a classroom setting with an instructor, you may be required to submit the maps you created in tutorial 5-5.

Turn in a printed map or screen capture of the following:

Tutorial 5-5.mxd

Two maps from Exercise 5-5.mxd

Tutorial 5-5 review

The multiple-ring buffers are similar to the other buffers in that they are measured on a straight-line distance, and they create a feature whose area can be measured. They also provide a dissolve option similar to the other buffers that can keep the buffer areas separate, or cause them to overlap as necessary.

Using any of the buffer tools, the final stages of the analysis become a simple inside-outside analysis that you have learned about in the earlier tutorials. Once these buffer features are created, they can be used like any other feature for overlays, or to make more selections.

Study questions

1. Explain the process of using multiple-ring buffers to determine the exact area of a soil layer that falls within each ring.

2. Can the same multiple-ring buffer process be performed with the regular Buffer tool or with the straight-line selection?

Other real-world examples

The local international airport might create multiple-ring buffers around its runways. These buffers can be used to set noise mitigation zones that can determine the amount of sound insulation a nearby house must contain.

The fire department creates multiple-ring buffers on the fly for large fires. These buffers are used to set up restrictive zones to control who can enter or leave each zone.

5-1
5-2
5-3
5-4
5-5
5-6
5-7
5-8
5-9

Tutorial 5-6

Quantifying nearness

There are other common methods for finding what is nearby. These methods include the Near analysis, which calculates the straight-line distance from specific features to their nearest neighboring features, and spider diagrams, which visually display the distance from a specified feature to all the other features in the dataset.

Learning objectives

- *Calculate straight-line distance*
- *Draw spider diagrams*
- *Perform distance analysis*

Preparation

- *This tutorial requires a custom script to perform the spider diagram analysis. It is provided in the \GIST2 \Toolboxes folder.*
- *Read pages 129–31 in* The Esri Guide to GIS Analysis, *volume 1.*

Introduction

Many of the tools introduced so far only indicate whether a feature is nearer than or farther than a given distance from a feature. They do not answer the question of exactly how far it is from the feature. However, you will learn two processes that help quantify nearness.

The first process is the Near analysis. A point is given as the central feature, and a distance to each of the other features is calculated. The distances are stored in the attribute table, and analysis can be performed using an exact distance quantity for each feature. The features may be color-coded by the distance from the source, or used to identify the closest source.

The second process uses the spider diagram generator. Similar to a Near analysis, a central point is given, but this time a line is generated from that point to every other feature. Because ArcMap automatically gives lines a Shape_Length field, the distance is calculated, too. Straight-line travel distance can easily be displayed for visual analysis.

Scenario The fire chief has provided some response data from last month and wants you to determine how many calls were answered by a crew from a station that was not the closest to the scene. From this information, the chief may be able to spot some understaffing or problems with the dispatch system.

Data   The chief provided the fire response data. It includes many fields, including the addresses used to geocode the locations. You will use the Station field to determine which station responded to each call.

The tutorial also uses the Create Spider Diagrams tool, written by Anthony Palmer of the US Army Corps of Engineers. It allows many varieties of spider diagrams to be created from the single script.

Use the Near tool

1   In ArcMap, open Tutorial 5-6.mxd.

The map displays the locations of all the fire calls that month. Notice that some calls were outside of the city limits. Oleander has a mutual-aid response agreement with all the neighboring cities to provide emergency service, regardless of city limits.

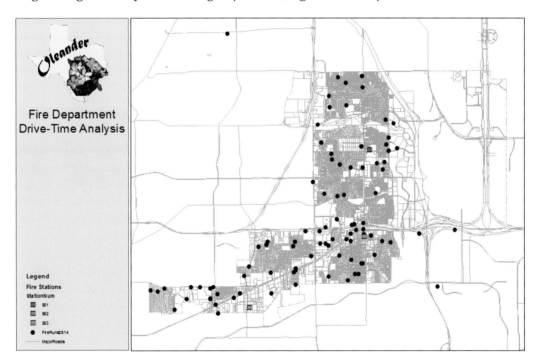

5-1
5-2
5-3
5-4
5-5
5-6
5-7
5-8
5-9

2   Use the Search window to find and open the Near tool.

The method of analysis will be to run the Near tool and determine which station is closest for each call location. Then you will match that number to the actual responding station to determine which stations had responders traveling too far.

3 Click the Near tool to run it. Use the drop-down list to set Input Features to FireRuns0514. Set Near Features to Fire Stations. When your dialog box matches the graphic, click OK to run the tool.

4 Right-click the FireRuns0514 layer and click Open Attribute Table. Move the slider all the way to the right to see the new fields that are added to the table.

The NEAR_FID and NEAR_DIST attributes are added to the table, identifying which station is nearest and the straight-line distance to that station, respectively. The Feature ID from the Fire Stations table was transferred to the FireRuns0514 layer, and you will join these tables to determine the station number for comparison.

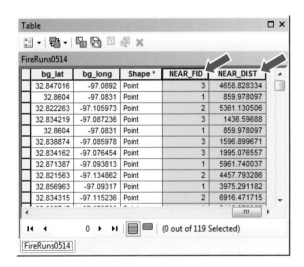

Join tables

1 Close the attribute table. Right-click the FireRuns0514 layer and click Joins and Relates > Join.

2 Set the Join type to "Join attributes from a table."

3 Set the Join field to NEAR_FID, which the Near tool created.

4 Set the Join table to Fire Stations.

5 Set the field in the join table to OBJECTID. When your dialog box matches the graphic, click Validate Join. A list of validation checks for the join are displayed.

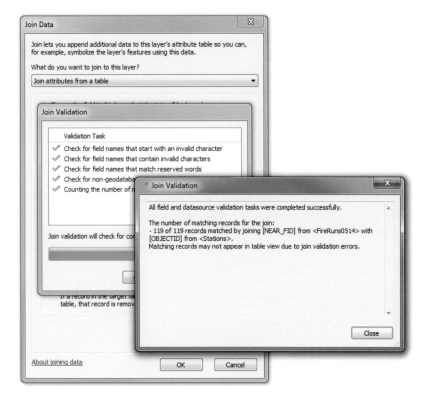

5-1
5-2
5-3
5-4
5-5
5-6
5-7
5-8
5-9

6 If the join passes the validation, close Join Validation and click OK to perform the join. Otherwise, recheck your settings and validate the join again.

7 Open the attribute table of the FireRuns0514 layer again and scroll to the far right.

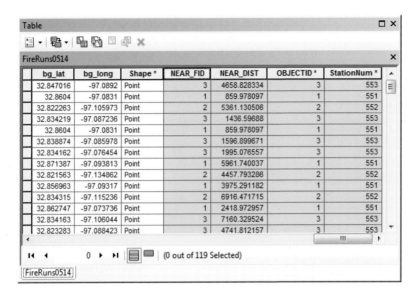

The Join procedure has added the attributes from the Fire Stations layer to the FireRuns0514 layer. Now you can do some checking to see the cases in which the responding station was not the nearest station. You will check by building a definition query to look for features for which the station numbers are not equal.

8 Close the attribute table.

If you remember from the previous tutorials, a definition query is used to limit the data that is shown in a feature class or table. Although the query does not delete any data, it makes subsets of the data invisible to the user.

9   Right-click the FireRuns0514 layer and open the properties. Go to the Definition Query tab and click the Query Builder button. Enter the query **FireRuns0514.STATION <> Stations.StationNum**. When your dialog box matches the graphic, click OK and then OK again to apply the query.

5-1
5-2
5-3
5-4
5-5
5-6
5-7
5-8
5-9

YOUR TURN

Set a unique value classification on the FireRuns0514 layer using the field STATION. Set the colors to match the existing colors for the Fire Stations layer, as noted. Adjust the symbol size as necessary.

       Station 551   Cretan Blue

       Station 552   Fire Red

       Station 553   Quetzel Green

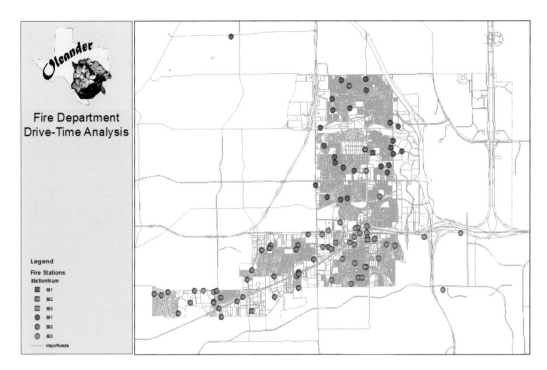

This map shows the calls that did not originate from the nearest station, and is ready for the fire chief to use for some quick visual analysis. The incidents may be investigated further to determine any nongeographic reasons for the problems.

One thing the map does not show is the other calls that correctly originated from the nearest station. Perhaps adding a spider diagram to the map will clarify the correct call responses.

10 Right-click the FireRuns0514 layer. Click Joins and Relates > Remove Joins > Stations to remove the join. Note that the definition query is automatically removed because it depends on fields from the join.

Create a spider diagram

A spider diagram draws a line from each response location to the fire station where the call originated. The mass of lines gives the viewer an overall pattern of distance and direction. The tool to create a spider diagram is available on the ArcScripts website, or in the data included with this book on the book resource web page.

1 Open the Catalog window 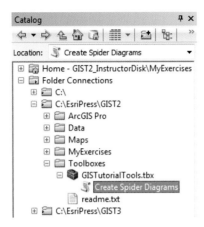 and locate GISTutorialTools.tbx in the Toolboxes folder. Double-click Create Spider Diagrams.

2 Click the down arrow for Origin Feature Class and select FireRuns0514. Set Origin Key Field to STATION.

3 Set Destination Feature Class to Fire Stations. Set Destination Key Field to StationNum.

4 Finally, set Output Feature Class to \GIST2\MyExercises\MyData.gdb, and name the feature class **FireSpider**. When your dialog box matches the graphic, click OK to start the process.

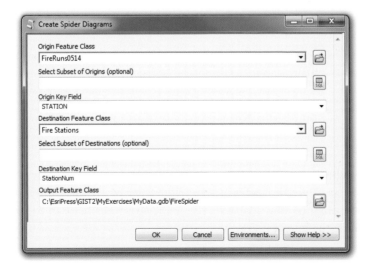

5-1
5-2
5-3
5-4
5-5
5-6
5-7
5-8
5-9

The Create Spider Diagrams tool draws a line from every feature in the FireRuns0514 layer to each fire station. To get the correct view of the data, look only at the lines where the ORIG_ID and DEST_ID are the same. The lines for which these two values are not equal do not represent real calls, but were made in the blanket operation of the Create Spider Diagrams tool.

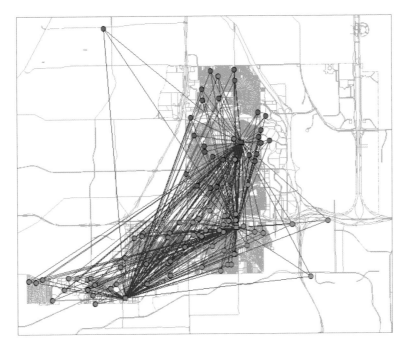

To restrict the data, you will set up a definition query in which ORG_ID equals DES_ID. This query will match the calls to their originating station.

5 Right-click the FireSpider layer and open the properties. Go to the Definition Query tab and build the query ORG_ID = DES_ID. Click OK and then OK again to accept the query.

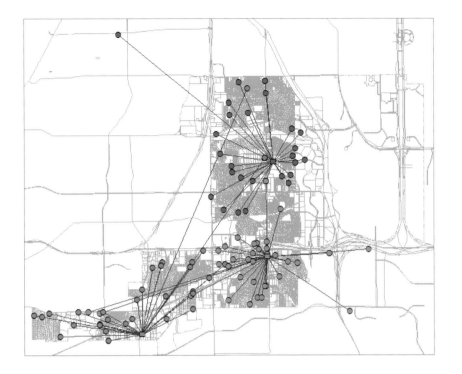

5-1
5-2
5-3
5-4
5-5
5-6
5-7
5-8
5-9

YOUR TURN

Set the symbology of the new FireSpider layer to unique values using DES_ID as the Value field. Color each value to match the color of the station symbols they are linked to. Adjust the line thickness as necessary.

The final map shows all the calls, along with a spider diagram color-coded to match the responding station. You are giving the fire chief a little more than asked for, but this will provide more information for visual analysis.

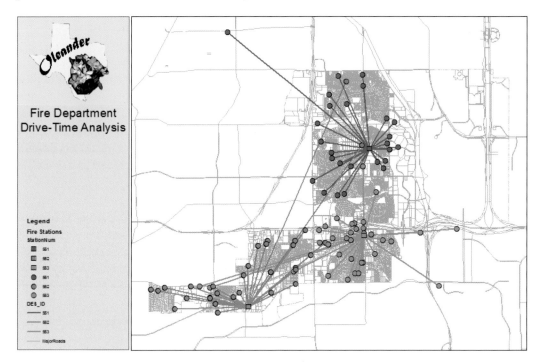

6   Save your map document as **Tutorial 5-6.mxd** to the \GIST2\MyExercises folder. If you are not continuing to the exercise, exit ArcMap.

Exercise 5-6

The tutorial showed how to find the nearest feature to a set of input points. The Near tool calculates the nearest feature and stores the distance to that feature. In addition, the spider diagram tool creates a linear feature from each point to its nearest point.

In this exercise, you will repeat the process using a different dataset. The tutorial used the Fire Department calls for service, showing the relationship between the call location and the fire station from which the call was dispatched. For this exercise, you will create a similar map using ambulance calls and analyze the relationship between the call locations and the fire stations from which the ambulances were dispatched.

- Continue using the map document you created in this tutorial, or open Tutorial 5-6.mxd from the \GIST2\Maps folder.
- Turn off the FireRuns0514 and FireSpider layers.
- Add the AmbulanceRuns0514 layer from the \GIST2\Data\City Of Oleander \FireDepartment feature dataset.
- Run the Create Spider Diagrams tool to create the same analysis as you did with the fire run data.
- Set all the symbology for the layers, as well as any other title changes or notes, to make the map visually appealing.
- Save the results as **Exercise 5-6.mxd** to the \GIST2\MyExercises folder.

What to turn in

If you are working in a classroom setting with an instructor, you may be required to submit the maps you created in tutorial 5-6.

Turn in a printed map or screen capture of the following:
> **Tutorial 5-6.mxd**
> **Exercise 5-6.mxd**

5-1
5-2
5-3
5-4
5-5
5-6
5-7
5-8
5-9

Tutorial 5-6 review

The Near function can be used to quantify a discrete feature's proximity to another location. With buffers, all you knew was that a certain location was farther away than one distance, but closer than another. With Near, an exact distance is recorded. It is important to note, however, that a point must exist at the location for which you want the distance value. Using the Near tool with line and polygon features will not give as precise a result because the measurements for these features are made to their centroids.

The spider diagram tool is almost always used for flashy visuals. The distance for each spider line is calculated, but because they are all straight-line distances their usefulness is limited. Leaving all the spider lines visible provides a good graphic display of the total service area for the data. The fire department data you worked with gave you the ability to separate each station and link them to only the calls they made, giving you three service areas, one for each station.

Study questions

1. When do you show spider lines for all origins and all destinations, and when do you separate them?

2. What is the major drawback of the Near tool?

Other real-world examples

A spider diagram might be created between known bird nests and sightings of an endangered species. This represents the travel territory of the birds.

The Near command can be used with business locations and public transit stops to determine which stop is closest, next closest, and so on.

Insurance companies may use the Near value of a house to fire hydrants to help set rates.

Tutorial 5-7

Creating distance surfaces

Straight-line distance analysis can also be performed using raster data. The values are formed into a continuous coverage, in which each pixel contains the distance from the subject feature. The process can include a cost to traverse the distances, resulting in a cost-distance surface.

Learning objectives

- *Work with raster data*
- *Calculate costs*
- *Perform distance analysis*

Preparation

- *This tutorial requires the ArcGIS Spatial Analyst extension.*
- *Read pages 132–34 and 142–47 in The Esri Guide to GIS Analysis, volume 1.*

Introduction

The Spatial Analyst extension allows for many of the same types of analysis that you perform with vectors to be done on raster images. The advantage is the ability to overlay many layers in one process, and to have the process complete quickly.

The distances you calculated for the discrete point data were only for those points. Any area that does not have a point does not have a distance value calculated. With a raster image file, the data is continuous across the entire study. So even if there are not discrete points at a location, a distance value is stored, creating a continuous distance surface. In raster analysis, it is possible to overlay this distance surface with data such as slope to find areas that have a certain slope and distance value combined.

The distance surface can also include a cost to traverse the surface. The inclusion of this cost adds another factor to the analysis, allowing you to investigate other influences on the nearness of features.

Scenario Using the data from the fire chief, you will create a distance surface that can be used as a backdrop for some of the data analysis. You can do a smooth gradation of color over the entire distance, which might add some impact to your map.

Data The Fire Stations layer will be used to set the start locations for the distance surface. The fire calls data will allow you to perform some quick visual analysis for distance.

Create a raster ring buffer

1 In ArcMap, open Tutorial 5-7.mxd.

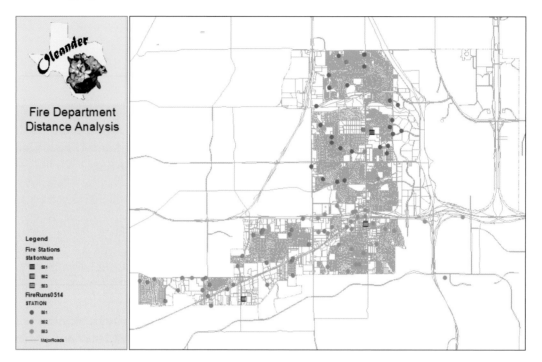

Here are the fire calls and station locations again. You will perform the distance analysis for Station 1 by making a single distance surface that goes out three miles. Next, you will do the same for Station 2 and see what areas are within one mile of both stations.

2 On the main menu, click Customize > Extensions. If necessary, select the Spatial Analyst check box. If this extension does not appear in your list, or an error message responds that Spatial Analyst is not available, see your license administrator.

3 Check to make sure that Fire Station 1 is selected. If it is not, use the Feature Selection tool to select it.

You will also set a processing extent. By default, Spatial Analyst uses the extent of the input file, which almost always works for raster files. But your input file is the point locations of the fire stations, which means that the default extent is the smallest box necessary to enclose all the points. The result will not cover the entire city, so you will set an extent in the geoprocessing environment that does cover the entire city.

4 On the main menu, click Geoprocessing > Environments. Click the Processing Extent arrow to open the input dialog box. Change Extent to Same as layer Lot Boundaries. Click OK.

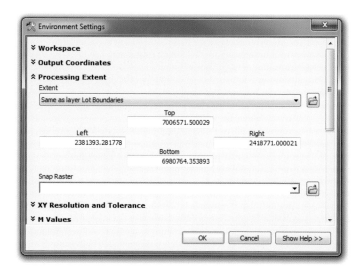

5 Open the Search window and locate the **Euclidean Distance** tool. Click to open it.

6 Use the down arrow to set Input raster or feature source data to Fire Stations. Set Output distance raster to MyData.gdb and name it **Station1_Distance**. Set Maximum distance to **10560** (two miles) and Output cell size to **50**. When your dialog box matches the graphic, click OK.

The result is a bull's-eye layer highlighting the distance from the fire station. The default classification uses only a few colors, making the rings look a little rough. You will smooth that out later when the final version is done.

5-1
5-2
5-3
5-4
5-5
5-6
5-7
5-8
5-9

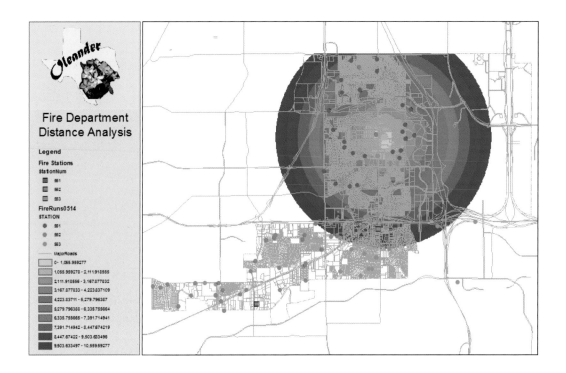

YOUR TURN

Clear the selected features and select Fire Station 3 (the green symbol). Then repeat the process. Make a distance surface using the distance of **10,560** and a cell size of **50**. Save it as **Station3_Distance**.

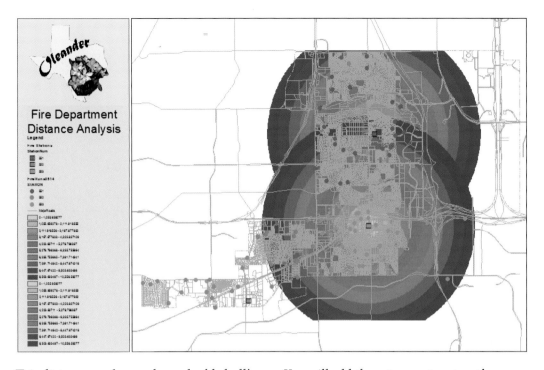

This distance surface makes a double bull's-eye. You will add these two rasters together, which means that at each pixel the values for each layer will be added. Locations with a close proximity to both stations will get a higher value, and those farther away will get a lower value. You will use the Spatial Analyst tool Cell Statistics to add the cell values together.

Combine rasters

1   In the Search window, click the **Cell Statistics** tool to open it.

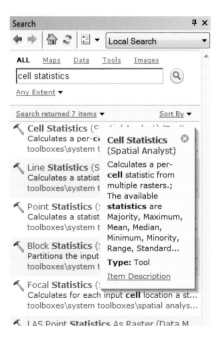

5-1
5-2
5-3
5-4
5-5
5-6
5-7
5-8
5-9

2 Set the inputs as both Station1_Distance and Station3_Distance. Make the output file **Station1_Plus_3_Distance** in MyData.gdb. Set the Overlay statistic to SUM; and if necessary, clear the "Ignore no data in calculations" check box. When your screen matches the graphic, click OK.

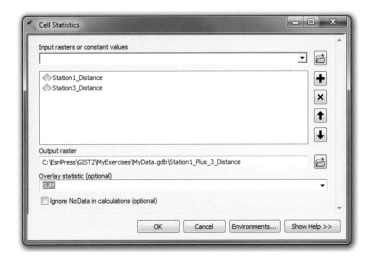

3 Turn off the two bull's-eye layers you created before, leaving the new layer visible.

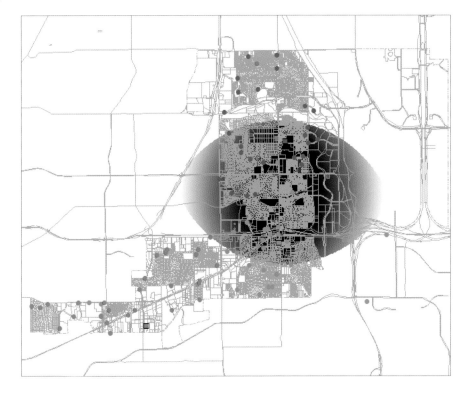

The default classification is a grayscale with the lowest values as the darkest. You will change this grayscale to a red palette with the highest values as the darkest.

4 Right-click the Station1_Plus_3_Distance layer and open the properties. On the Symbology tab, click the Color Ramp down arrow and select a monochrome red scale with the darkest value on the right. Click OK.

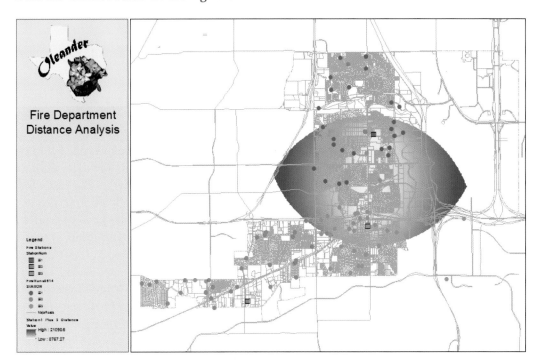

The red zone shows relative distance from each fire station. The darker areas are farther away, making them more of a concern for the fire chief.

5 Save your map document as **Tutorial 5-7.mxd** to the \GIST2\MyExercises folder. If you are not continuing to the exercise, exit ArcMap.

Exercise 5-7

The tutorial showed how to create distance surfaces, and then combine the results into a single raster by doing a cell-by-cell summation. The result is a continuous phenomenon dataset that displays distance from the fire stations.

Before you take this map to the fire chief, you will perform the surface distance analysis process for the intersections of Station 2 and Station 3.

- Continue using the map document created in this tutorial, or open Tutorial 5-7. mxd from the \GIST2\Maps folder.
- Create a distance surface for Station 2 (and for Station 3 if you did not do it in the tutorial).
- Add the surface for Station 2 to the surface for Station 3 using the Cell Statistics tool.
- Adjust the legend and symbology for the new surface.
- Save the results as **Exercise 5-7.mxd** to the \GIST2\MyExercises folder.

What to turn in

If you are working in a classroom setting with an instructor, you may be required to submit the maps you created in tutorial 5-7.

Turn in a printed map or screen capture of the following:

Tutorial 5-7.mxd
Exercise 5-7.mxd

Tutorial 5-7 review

The distance surface commands are practically the same as the Near command, except that a value is calculated for every location on the map. Using the Identify tool, you can click anywhere on a raster surface and get a distance value.

The distance surface also requires you to perform two steps to get a similar result as the Near command. First, you can create a distance surface for only one station at a time. Then you must add the two rasters together to get a value that represents the distance to the nearest fire station.

The distance surface also suffers from the same major drawback as the Near command: the distances are all straight line (Euclidean). Things such as lakes and freeways are not considered as obstacles in getting from one place to another.

Study questions

1. How are the raster values different from the point values that were created using the Near command?

2. What data is stored in a density surface?

3. When are the raster surfaces created with the Euclidean Distance tool not appropriate for distance analysis?

Other real-world examples

A radio station may create a distance surface to determine a reception area. The overlapping areas might be used to investigate the impact of the radio waves on animal habitat.

In performing raster analysis, the distance surface can be used to introduce a distance element to the formula. A distance surface may be generated around a body of water and added to the raster data for crop health to investigate possible contamination sources.

5-1
5-2
5-3
5-4
5-5
5-6
5-7
5-8
5-9

Tutorial 5-8

Calculating cost along a network

Analysis using straight-line distance has an inherent flaw when dealing with streets or other types of networks. The distance measurements may cross areas in which the network does not go. The solution is to use ArcGIS Network Analyst tools to calculate a cost along the network.

Learning objectives

- *Build a network*
- *Calculate costs*
- *Perform distance analysis*

Preparation

- *This tutorial requires the ArcGIS Network Analyst extension.*
- *Read pages 135–41 in* The Esri Guide to GIS Analysis, *volume 1.*

Introduction

All the tutorials you have done with fire stations so far have involved measuring a straight-line distance (Euclidean distance, or "as the crow flies") from the analysis location points. The problem with this method is that sometimes there are places where the streets do not go, such as across lakes or other natural barriers.

The Network Analyst tools allow you to overcome some of these difficulties by performing distance analysis along the network. Networks can be streets, pipelines, electric transmission lines, creeks, train tracks, and so on. These networks all have connectivity, and something traverses them. To get the most accurate analysis, assign a cost for traversing each segment of the network. If you have cost per mile for each street segment, the Network Analyst extension can show how far you can go for a dollar. Similarly, electric transmission lines might be limited by capacity in volts, or a pipeline might be restricted by flow rate based on pipe size.

For this tutorial, you will calculate the time it takes to traverse each segment and see how far you can go in a specified number of minutes. This measurement is more accurate than a distance-only cost because it takes into account speed limits. Note that although you are using time as the cost, the process will work equally well with distance only, if speed limit data is not available.

Scenario   It seems that anytime you do a great map, the fire chief suddenly thinks of more maps to request. After analyzing the previous maps, the fire chief has seen that Station 3 crews are covering Station 1 quite a lot. It seems that the new development at the north end of town is calling out the Station 1 crew more often, leaving other crews to cover the rest of its district. What you cannot see in your analysis is the number of times that nearby cities backed up Station 1 and came into Oleander on fire calls under the mutual-aid agreement.

The chief wants to start the process with the city council to approve a new station. Part of the justification is to show that there are areas that are not being covered in the standard response time. You will use the Network Analyst extension to show how far the fire trucks can go in a given amount of time.

Data   The street centerline file will be used for the network analysis. The fields SpeedLimit and Shape_Length are used to calculate the time it takes to traverse each line segment.

The Fire Stations layer is used to set the start locations for the analysis.

Create a network database

1   Start ArcMap, open Tutorial 5-8.mxd, and activate the Network Analyst extension. Click the Catalog tab on the right of the map document, navigate to the directory where you installed the tutorial data, and expand the Data folder. Right-click Networks. gdb and click Copy.

2   Scroll down in the Catalog tree and find the MyExercises folder. Right-click it and click Paste.

This gives you a copy of the data to use for the network, in case things get messy. It will also provide you with a place to store all the analysis results.

You will add a field to the Street Centerlines feature class to store the time data you will calculate.

3   Expand the MyExercises folder, the Networks geodatabase, and the Network Analysis feature dataset. Right-click StreetCenterlines and click Properties.

5-1
5-2
5-3
5-4
5-5
5-6
5-7
5-8
5-9

4   In the Properties dialog box, go to the Fields tab. Scroll all the way to the bottom of the fields list and click in the first open line. Type a new field name—**Minutes**—and set Data Type to Double. Then click OK.

You will create a network using this new field as a cost, which will allow you to perform routing based on the number of minutes it takes to complete the route. The field is empty now, but you will populate it with values later.

5 Right-click the feature dataset Network Analysis and click New > Network Dataset to start the creation process.

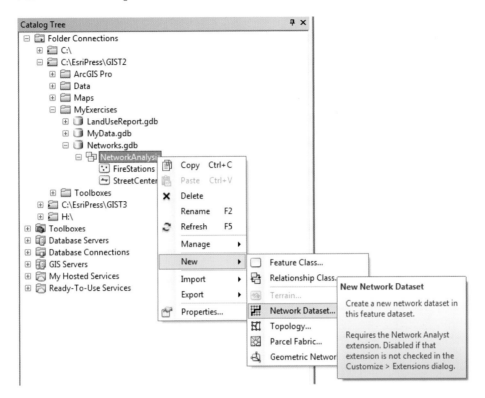

6 In the first screen of the Network wizard, name the new dataset **FireDriveTime**. Note that if you intend to use this network with earlier versions of ArcGIS software, be sure to select the appropriate version.

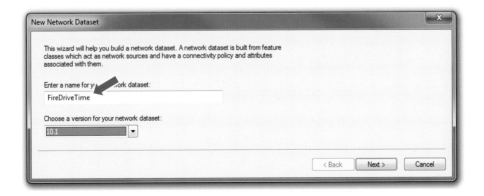

5-1
5-2
5-3
5-4
5-5
5-6
5-7
5-8
5-9

7   Click Next. Select the check box to add StreetCenterlines to the network dataset.

8   Click Next. Accept the defaults for the next seven screens, clicking Finish to create the network dataset.

The important screen in the process is to set your new field as the cost. But because you used the default keyword of Minutes, it is automatically used for the cost field. Other keywords that are recognized as a cost for a network can be found in the file NetworkDatabaseConfiguration.xml, which loads with the Network Analyst extension.

9   When prompted to build the network, click No. ArcMap automatically adds the network dataset to your map document. Click Yes when prompted to add participating features classes. You will build the network after the Minutes field is populated.

Set up the network

1   If necessary, close the Catalog window to reveal the map.

The map looks similar to the ones in the previous tutorials that display the Fire Department calls for service and the street centerlines. You will add the new street network you built, calculate the Minutes field for each line segment, and identify the fire stations as the start points for the analysis.

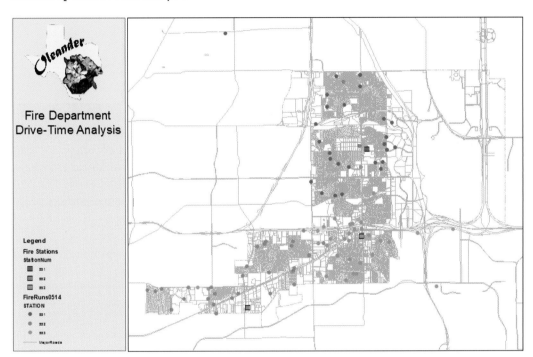

5-1
5-2
5-3
5-4
5-5
5-6
5-7
5-8
5-9

2   Right-click the StreetCenterlines layer and click Open Attribute Table. Scroll the table all the way to the right and find the Minutes field.

3 Right-click the Minutes field and click Field Calculator. If necessary, start an edit session and click Continue on the warning box. Build the equation as follows:

[Shape_Length] / (([SPEEDLIMIT] * 5280) / 60).

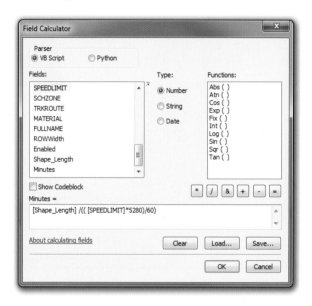

4 Click OK. The result is the time in minutes that it takes to traverse each line segment. Close the attribute table.

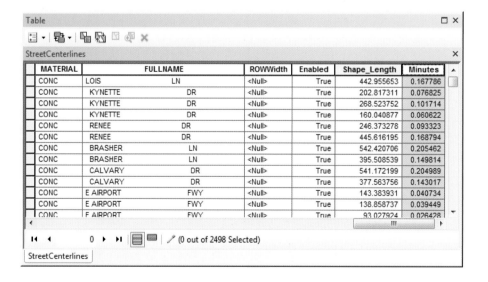

Perform network analysis

1   If necessary, turn on the Network Analyst extension in ArcMap. Add the Network Analyst toolbar to ArcMap by clicking Customize > Toolbars > Network Analyst on the main menu. Dock it if you want.

By default, the toolbar displays the FireDriveTime network because it is the only one that exists in the table of contents. Now that you have populated the Minutes field with the time it takes to traverse each line segment, you can build the network.

2   Click the Build Network Dataset button 🌐 .

This network will now allow you to perform several types of network analysis, including routing, finding the nearest facility, creating an origin–destination cost matrix, and the process you are going to use: working with a service area.

The Closest Facility function is similar to the Near tool that you used earlier, except that it measures distance along the road instead of the straight-line distance. You will use this function in tutorial 5-9. The Origin–Destination (OD) cost matrix function produces similar results as the Spider Diagram tool you used earlier.

The first step is to add the framework in which your analysis will take place. To add the framework, you will use the New Service Area tool.

3   On the Network Analyst toolbar, click Network Analyst and click New Service Area.

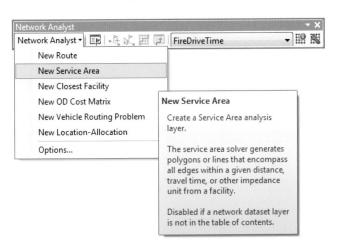

5-1
5-2
5-3
5-4
5-5
5-6
5-7
5-8
5-9

This tool adds a new group layer to the table of contents named Service Area, along with a series of feature classes that will hold the components of your analysis. These feature classes are currently empty, but they will be populated as you work with the network.

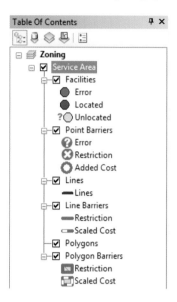

The Facilities layer will hold the start locations for your service area analysis. These locations will be the fire stations. You can place them into the feature class manually using the Create Network Location tool on the Network Analyst toolbar, which is not too bad for the three stations. But if you have more stations, you will need an automated tool. You will go ahead and use the automated tool to add the locations.

4 Use the Search window to locate and open the Add Locations tool.

Even though you have the fire stations as a feature class in your map, they must be matched up with the network database that you created. The Add Locations tool will move them into the framework of the network so that they can be used in the analysis.

5 In the drop-down list, set Input Network Analysis Layer to Service Area and set Sub Layer to Facilities.

6 Set Input Locations to Fire Stations. In the Field Mappings pane, set the Name field to **StationNum**.

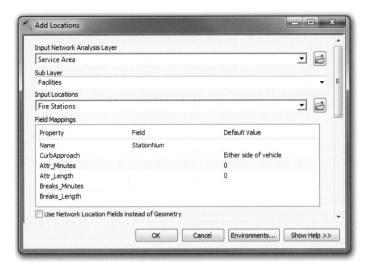

7 Scroll to the bottom of the dialog box and select the Snap to Network check box. When your dialog box matches the graphic, click OK.

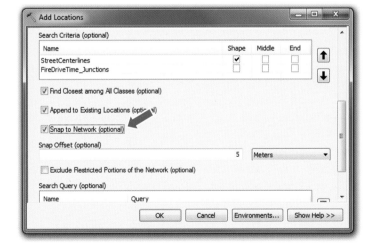

If you zoomed to the fire stations, you will notice a new dot placed there by the Add Locations tool. This point is now part of the network database and can be used with the various network tools. If you noticed, there are three other sublayers called Barriers to store point, line, and polygon features. You can place these elements on the street network to prevent the analysis from going through that point. For instance, if a road was closed for repair, you can place a barrier at either end, and that road segment will not be used in routing calculations.

5-1
5-2
5-3
5-4
5-5
5-6
5-7
5-8
5-9

Now that you have the fire stations added as network facilities, you can set the parameters for the service area calculation. You will define the measurement units you are going to use and the distance from the fire stations you want your analysis to extend.

The fire chief reports that he wants to see service areas for total response times of three, four, and five minutes. The standard emergency response allows 60 seconds to dispatch the call to a station, and another 60 seconds to get the firefighters on the trucks and out the door. Subtracting that two minutes from the total response times gives drive times of one, two, and three minutes to reach the location.

8   Right-click Service Area in the table of contents and click Properties.

9   In the Properties dialog box, go to the Analysis Settings tab. Impedance is already set to Minutes because ArcMap recognized the keyword as a valid field name. Set Default Breaks to 1, to represent one minute. When your dialog box matches the graphic, click OK.

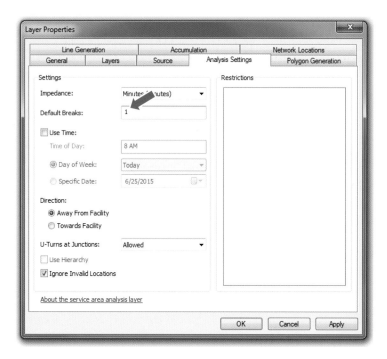

Everything is set up for the analysis: the facilities are loaded, and the service area parameters are set to 1 minute.

10   On the Network Analyst toolbar, click the Solve button 🔲 .

Polygons are created showing the service areas for each station. Notice also that they are not the nice, rounded buffers you got before. This is a good example of the difference between straight-line distance and distance along a network.

11 Turn off the street centerlines and FireDriveTime_Junctions layers to make your map easier to read.

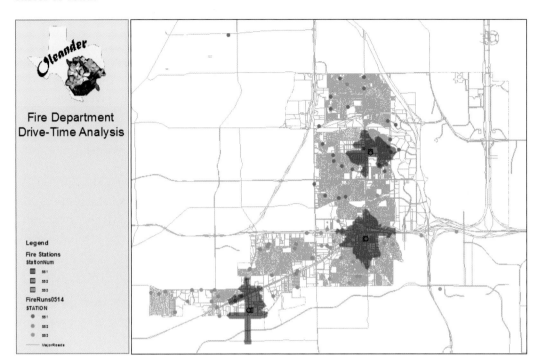

Export results

The polygons created in the drive-time analysis are temporary. If you change the parameters and solve for service areas again, they will disappear so you will export the results to a new feature class.

1 In the table of contents, right-click the Polygons layer and click Data > Export Data.

2 Click the Browse button and set Output Feature Class to **Service_Areas_1_Minute** in \MyExercises\Networks.gdb. Click Save and click OK. When prompted, elect to add the new layer to the map document.

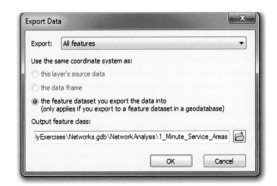

YOUR TURN

Change the symbology of the Service_Areas_1_Minute layer to make it more presentable:

- Set a unique value classification using the Facility ID.
- Color-shade each facility according to the colors used previously for the stations.
- Set the transparency to **50** percent.
- Move the Service_Areas_1_Minute layer above Street Centerlines in the table of contents.

You may want to turn off the Service Area group layer to check your work, but turn it back on before proceeding.

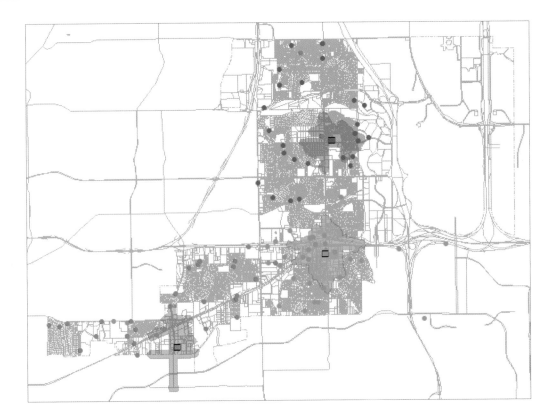

Now that you have mapped the one-minute service areas, you will also map the two-minute and three-minute service areas. Thus, you will change the parameters of the Service Area layer and solve again.

3   Turn off the one-minute service area layer. Right-click the Service Area layer and click
Properties. Go to the Analysis Settings tab and change Default Breaks to **2**. Click OK.

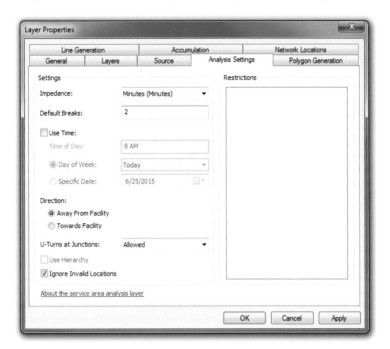

4   On the Network Analyst toolbar, click the Solve button.

5-1
5-2
5-3
5-4
5-5
5-6
5-7
5-8
5-9

The new service areas are generated, and you see much more coverage.

YOUR TURN

Export the Polygons 2 layer to a new feature class named **Service_Areas_2_Minute**, and add it to the map document.

Change the symbology of the new layer to make it more presentable, as follows:

- Set a unique value classification using the Facility ID and arrange the colors to match the one-minute drive time. (**Hint:** on the Symbology tab, click the Import button and use the settings from the one-minute layer.)
- Set the transparency to **60** percent, lighter than the other layer.
- Repeat the whole process to create the three-minute service areas.
- Set the colors as before and set the transparency to **80** percent.
- Review the order of layers, and move them in the table of contents to make the map clear and concise.
- Turn off the Service Area layer, which will turn off all its associated layers.

The map now shows the one-, two-, and three-minute response zones, along with a few areas that do not seem to have good coverage. Perhaps this map will help the case for funding a new fire station.

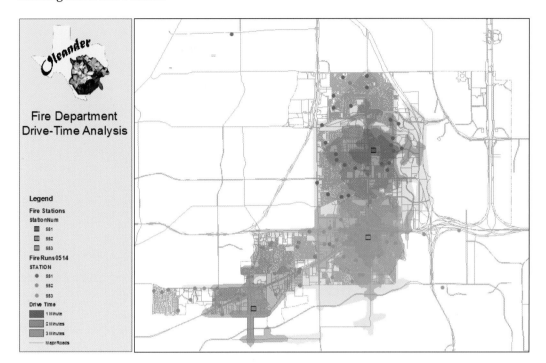

In this tutorial, you created each of the response zones separately. However, you can also do it in one step. The Analysis Settings dialog box accepts multiple costs separated by a space, and other settings allow you to decide between overlapping polygons and ring polygons. Explore these options to see how they can be used for future projects.

5 Save your map document as **Tutorial 5-8.mxd** to the \GIST2\MyExercises folder. If you are not continuing to the exercise, exit ArcMap.

5-1
5-2
5-3
5-4
5-5
5-6
5-7
5-8
5-9

Exercise 5-8

The tutorial showed how to establish a service area, based on the cost of traveling through a network. The "cost" is time, and the "network" is the streets.

The fire chief saw the map, was impressed, and then asked if you can count the number of buildings in each of the three service areas for each time value. The results of the service area analysis are polygons, and with building footprints available, it becomes a simple "finding what's inside" task. The chief will want the map when it is finished.

- Continue using the map document you created in this tutorial, or open Tutorial 5-8.mxd from the \GIST2\MyExercises folder.
- Add the layer BuildingFootprints from the City of Oleander geodatabase, Planimetric_Data feature dataset.
- Use each of the three service areas to determine the number of buildings in each.
- Add a text box to the map title area to show the results.
- Set all the symbology for the layers, as well as any other title changes or notes, to make the map visually appealing.
- Save the results as **Exercise 5-8.mxd** to the \GIST2\MyExercises folder.

What to turn in

If you are working in a classroom setting with an instructor, you may be required to submit the maps you created in tutorial 5-8.

Turn in a printed map or screen capture of the following:
> **Tutorial 5-8.mxd**
> **Exercise 5-8.mxd**

Tutorial 5-8 review

Finally, the problem of straight-line distance analysis is solved. Measurement along the streets is a better representation of travel time than the "as the crow flies" model of the buffer and raster surface tools.

It took a lot of data setup to get the Network Analysis tools to work. A good street centerline file with perfect connectivity is needed, along with a way to calculate some cost along the line. In many cases, the cost is merely distance, but using only the distance as a cost does not give priority to freeways on which you can drive much faster. Your dataset has speed limits, which allows you to calculate the traverse time for each line segment. It is also important to be aware of one-way streets and restricted turns. Your database does not account for these rules, but a more complex dataset may have these characteristics built in.

Remember that a network is only a model of reality. You can never build in the true complexity of a real street network. Why don't the Internet mapping programs produce the exact route you take to work every day? Because your observational model is more complex. You know when traffic will be heavy, the condition of certain streets, and the reality of making some of the more difficult turns. The computer models are much more generalized, although the ability exists to make them more complex.

Study questions

1. At what point does the accuracy of your network model negatively affect your results?

2. How can a model based on speed limits be more accurate?

3. You performed only drive-time analysis on this model, although it is capable of routing. Using the Network Analyst toolbar, try adding some stops and an origin. Are the routes created by the Network Analyst tools acceptable?

Other real-world examples

A flower delivery service might create a drive-time polygon around each of its member stores to show what areas are serviceable, and which store will deliver to a particular area.

The newspaper office may use routing to determine the best path for its delivery agents, provided a start point and a list of delivery locations.

The police department's criminal investigation unit might create a drive-time polygon to see if a suspect can drive from his location to the crime scene in 15 minutes.

5-1
5-2
5-3
5-4
5-5
5-6
5-7
5-8
5-9

Tutorial 5-9

Calculating nearness along a network

The Near command from the general ArcToolbox analysis toolset uses a straight-line distance calculation, which is not always the most accurate method. Using Network Analyst, you can duplicate the Near command, and accurately measure the results along the network.

Learning objectives

- *Calculate proximity along a network*
- *Perform distance analysis*

Preparation

- *This tutorial requires the ArcGIS Network Analyst extension.*
- *Review pages 129–31 in* The Esri Guide to GIS Analysis, *volume 1.*

Introduction

You solved the "as the crow flies" dilemma by building a network and using it for analysis. Another tool in Network Analyst mimics the Near tool you used in tutorial 5-6. The Near tool measures the straight-line distance from a source location to each feature in the input dataset. The Closest Facility tool runs the same analysis but gives the results as the cost along a network.

As stated previously, using this tool mitigates anomalies with the data such as having to drive around lakes or closed roads.

Scenario   Now that the fire chief knows your capabilities, he wants the Near analysis that you did earlier updated to follow the network. Several of the city council members remarked that the map had some calls going right through the high school football stadium, and that cannot be right. And although it is "right" for the tool you used, there is another way to calculate a near distance that is closer to reality.

You will repeat the analysis, but this time use the New Closest Facility tool in Network Analyst.

Data   The Networks database built in tutorial 5-8 will be loaded again, with minutes as the cost of traversing each road segment.

The Fire Stations and FireRuns0514 layers are used for the origins and destinations in the analysis.

Determine nearness using driving distance

1 In ArcMap, open Tutorial 5-9.mxd.

This map should look familiar, as it is the fire response calls and fire station locations. You will add the network database that you built in tutorial 5-8 to the map document; then you can start the analysis.

5-1
5-2
5-3
5-4
5-5
5-6
5-7
5-8
5-9

2 Click the Catalog tab and navigate to the FireDriveTime network you created in tutorial 5-8. Drag it to the top of the layer list in the table of contents. When prompted, add the other feature classes that participate in the network. Turn off the FireDriveTime _Junctions layer.

3 If your Network Analyst toolbar is not visible, add it now. On the Network Analyst toolbar, click Network Analyst and then click New Closest Facility.

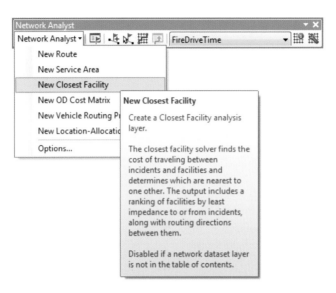

You have added a set of layers to contain the components of the analysis. You will add Facilities and Incidents to the layers as analysis locations. The Facilities are the fire stations, and the Incidents are the fire calls.

4 Use the Search window to locate and open the Add Locations tool.

5 Set Input Network Analysis Layer to Closest Facility and set Sub Layer to Facilities.

6 Set Input Locations to Fire Stations and the Name field to StationNum.

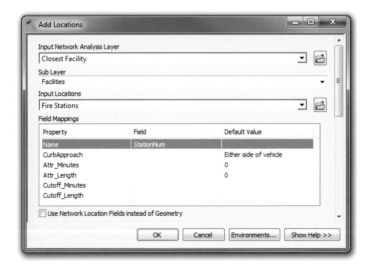

7   Scroll to the bottom of the dialog box and select the Snap to Network check box.

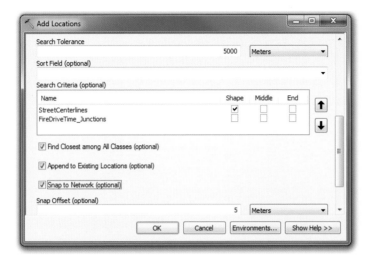

8   Click OK to run the tool.

YOUR TURN

Repeat the process to add the fire calls as incidents, as follows:

- Run the Add Locations tool.
- Set the input analysis layer.
- Set the sublayer to Incidents.
- Set the input locations to FireRuns0514.
- Set the tool to snap to network.

Everything should be set for the analysis. When you start the process, a route along the street network is calculated from each fire station to each response call.

5-1
5-2
5-3
5-4
5-5
5-6
5-7
5-8
5-9

9   On the Network Analyst toolbar, click the Solve button.

YOUR TURN

As in tutorial 5-8, the results of the analysis are temporary. You will save the routes to another feature class. Then you will symbolize the routes to represent the nearest station.

- Export the Routes layer as **NearNetwork** to \Networks.gdb\NetworkAnalysis under \MyExercises\MyData.
- Symbolize the new layer using the Facility ID field.
- Turn off the Closest Facility layer group, StreetCenterlines, and FireDriveTime.
- Review the order of layers and labels to make the map clear and concise.

This completed map should be just what the fire chief wanted. Now each route is shown in the color of the responding fire station. The situations in which the incident color does not match the route color indicate that the responding station was not the closest.

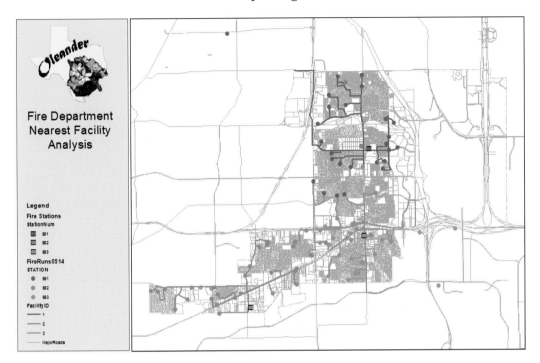

10 Save your map document as **Tutorial 5-9.mxd** to the \GIST2\MyExercises folder. If you are not continuing to the exercise, exit ArcMap.

5-1
5-2
5-3
5-4
5-5
5-6
5-7
5-8
5-9

Exercise 5-9

The tutorial showed how to use network routing to find the nearest facility to a set of incidents using time as the cost.

The fire chief also wants to see the results of the nearest facility along a network analysis for the ambulance calls. The process is the same as that used for the fire response calls, except for using this different dataset. Repeat the process using the ambulance response data you used in tutorial 5-6.

- Continue using the map document you created in this tutorial, or open Tutorial 5-9.mxd from the \GIST2\Maps folder.
- Add the layer AmbulanceRuns0514 from \GIST2\Data\City Of Oleander.gdb \FireDepartment.
- If necessary, add the Networks database for the streets that you created in this tutorial.
- Run the Closest Facility analysis.
- Set the fire stations as facilities and the ambulance calls as incidents.
- Set all the symbology for the layers to coordinate each set of calls to their originating station, and make any necessary title changes or notes to make the map visually appealing.
- Save the results as **Exercise 5-9.mxd** to the \GIST2\MyExercises folder.

What to turn in

If you are working in a classroom setting with an instructor, you may be required to submit the maps you created in tutorial 5-9.

Turn in a printed map or screen capture of the following:
> **Tutorial 5-9.mxd**
> **Exercise 5-9.mxd**

Tutorial 5-9 review

Calculating the nearest facility along a network was far more accurate than the straight-line distance calculations. Natural barriers such as rivers, as well as areas with no roads, now play an important role in the results. For the purposes of the fire chief, this method is much more realistic and valuable because the fire trucks are limited to driving on roads and do not handle cross-country driving well.

Again, a good road network must exist, and it is important that it have good connectivity. If such a database does not exist, it is possible to clean up an existing street centerline file to use for network analysis.

Study questions

1. What are the important differences between straight-line measurements and measurements along a network?

2. What are some of the requirements for a dataset to become a network database?

3. When does straight-line distance become a good model of reality?

Other real-world examples

A delivery company might want to use Network Analyst tools to determine which warehouse is the closest to a delivery site. Then the order can be dispatched using the most efficient route.

A public works department might build a network of the sewer lines to determine the nearest manhole to a reported problem. The distance can determine whether the problem is reachable by a remote device from more than one access point.

The Police Department may want to analyze call data as the Fire Department did to determine whether response times are in line with the distances the responding officers drove. A long response time to travel a short distance may point to problems in the dispatch system.

5-1
5-2
5-3
5-4
5-5
5-6
5-7
5-8
5-9

6
Mapping change

Temporal analysis deals with mapping change, which may occur as a change in location, a change in magnitude, or a change in one of the dataset's associated values. The change may be shown on a single map, or it may require a map series to show the results of the change. Features that change in more than one characteristic present an especially difficult challenge for cartographers. Carefully controlling the symbology and the amount of data being shown, as well as the number of maps used in a map series, help make it easier to present the results.

Tutorial 6-1

Mapping change in location

Datasets capture values at one instant in time and paint a picture of that event. But some events happen over time, so data is collected to show how the values change during a specific time event.

Learning objectives

- *Combine datasets*
- *Track an event*
- *Overlay data*

Preparation

- *Read pages 149–64 in* The Esri Guide to GIS Analysis, *volume 1.*

Introduction

Much of the data you work with represents a single location, at a single point in time. But data, like everything else, changes over time. So how do you map that change?

Items can change in one of three ways: by a change in location, a change in character, or a change in value. These changes may be displayed in a series of maps, in which viewers must form pictures in their mind of each map and do some comparison between them. More visual analysis! Another method is to overlay the data onto a single map, using transparencies to make the data readable. These overlays can also be used to make other selections and quantify some result of the change.

When mapping the change over time for analysis, you may use the data readings to predict a future event. For instance, mapping the recorded path of a hurricane helps predict where the storm may go next.

Mapping change over time can also show conditions before and after an event. Data is recorded in one snapshot in time. Then some event will take place, and then more data is collected after the event to determine how things have changed. Recording accident information before and after the installation of a new traffic control device might be used to show how effective the device is.

Mapping the time frame over which the change occurs is also a consideration in your analysis. The change may be recorded for specific dates or at a set interval. This interval can be an hour, a day, a week, a month, a year, and so on. The chosen time frame depends on how fast the data changes. Mapping daily house values, for example, is not effective. Perhaps that type of mapping is more of a monthly or annual thing. But mapping the expansion of a forest fire must be done within a short time frame.

When working with time-related data, you must also be aware of the duration, the number of values, and the interval of the study. The *duration* represents the total time over which the data is collected. The *number of values* is how many values are recorded over the duration, and the *interval* is the period between the time points when the values are recorded. An annual study with monthly values has a duration of one year, 12 values, and an interval of one month.

You can combine the collected data with other data to see how other features, in addition to the recorded features, change over time. Polygons that show the spread of a forest fire might be used to select the houses that have burned, giving a cost of destruction as the fire spreads.

Scenario   A large producer of book matches is located in Oleander, and the company stores a lot of flammable material on-site for match production. The fire chief wants to do a "tabletop" disaster drill in which a fire breaks out at the match plant. You are asked to show the movement of the chemical plume over the time frame of the drill.

Data   The main dataset is a set of polygons created with the freely available ALOHA plume modeling software to predict the progression of a plume because of winds, temperature, and the chemical agent involved. The software provides four predictions that are one hour apart.

A single layer that includes all the plume data as well as a time tracking field is also provided. This dataset will be used for a Time Slider series.

The building footprint data is derived from aerial photographs, and each building includes a field named UseCode that represents the building's use. The codes are as follows:
- 1 = Residential
- 2 = Multi-Family
- 3 = Commercial
- 4 = Industrial
- 5 = Government
- 6 = Utilities
- 7 = Schools
- 8 = Churches

Road and lot boundary layers are included for background interest.

6-1
6-2
6-3

Map the movement of a chemical plume

1 In ArcMap, open Tutorial 6-1.mxd.

The match factory has had an accidental ignition, and the resulting fire has caused a release of potassium chloride into the surrounding neighborhood. The first hour's plume is shown, known as the Level 1 response plume. The factory's safety measures are containing the plume, but the fire chief has provided some data from a plume-tracking software that will show where the plume may travel over the next three hours. You will use these predictive plumes to determine how many buildings to evacuate.

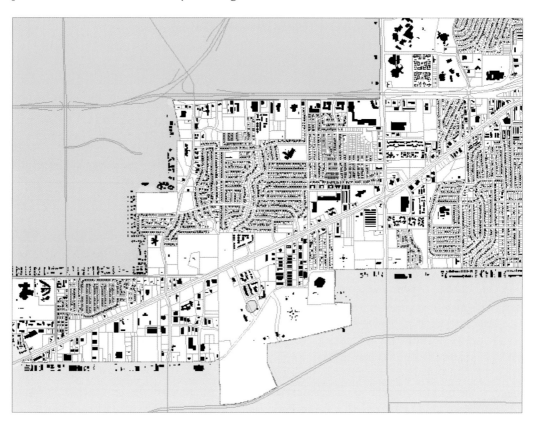

2 Open the properties of the Level 1 layer and go to the Symbology tab. Change Fill Color to Mango.

3 Click the Display tab and change the transparency to 20%. Click OK to see the results on the map.

Although you can do a quick visual analysis and see how many buildings are affected, for a better understanding you will use GIS tools to run through the process.

4 On the main menu, click Selection > Select by Location. Build the selection statement, as shown in the graphic. If the Building Footprints layer does not appear as an option, make sure that the layer is selectable. When your dialog box matches the graphic, click OK.

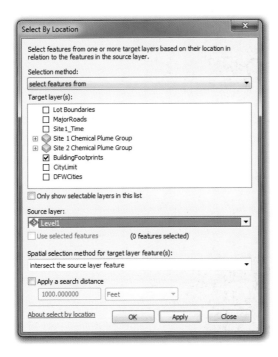

5 Open the attribute table of the BuildingFootprints layer. Make a note of how many buildings are selected and the use code.

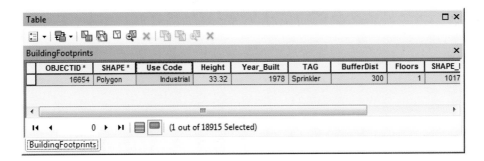

6 Close the attribute table. Clear the selected features. Add a text box to the map's title bar to show how many buildings of each use code are included in the Level 1 plume.

A single building is selected. For the purposes of the tabletop drill, the disaster area will be expanded and the response level elevated.

6-1
6-2
6-3

Note: remember to use the navigation tools on the Layout toolbar if you want to zoom in for a closer look.

The ALOHA plume-modeling software predicts that the plume will expand in one hour. The Level 2 layer represents this expansion. Next, try to determine what structures, and their uses, will be affected by this expanded plume.

7 Turn on Level 2. Open its properties and change the color to Fire Red and set the transparency to 50%.

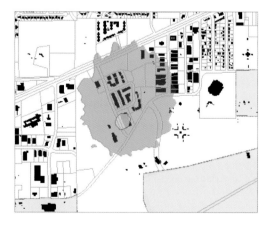

8 Use the Select by Location tool to select the buildings that are inside the second plume.

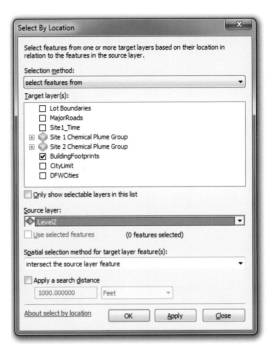

9 Perform a summary on the UseCode field to discover how many selected buildings belong to each use code. (**Hint:** see the graphic.) Add the results to the text box in the title block.

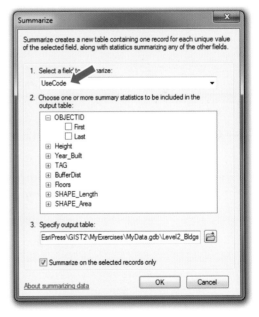

6-1

6-2

6-3

10 When you are finished, clear the selected features.

The numbers you derived from this analysis will help the Fire Department plan for this scenario. The fire chief will be able to determine the resources needed over what time frame and decide how to deploy these resources.

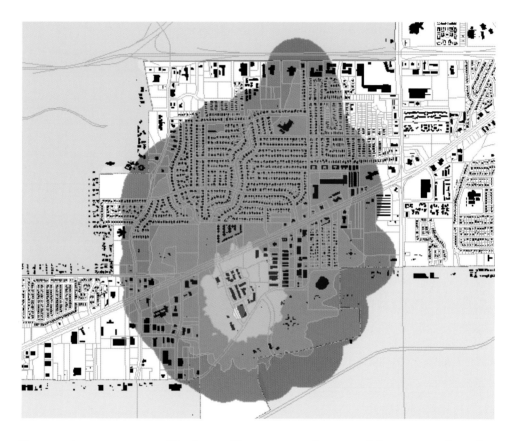

The exercise was a success. You learned how to map plumes and do some quick analysis; and now, the Fire Department can visualize the impact of the plume over a four-hour duration.

Time Slider analysis

If you want a visual representation of how the spread of the chemical fallout will grow over the four hours, the Time Slider tool will allow you to set up the means to track it using time tracking. The layer for the time series must contain all the data and have a time field for each iteration that you want to show. The iterations are set up on the Time tab of the layer properties and then viewed using the Time Slider.

1   Turn off the Site1 Chemical Plume Group. Using the Catalog tab, navigate to the
    \GIST2\Data folder. Drag the Site1_Time layer to the table of contents and place it
    directly above the Site 1 Chemical Plume Group. The classification value field is the
    time of occurrence.

    The symbology is set to reflect time. Note that the event started at 5 a.m. and was measured
    once an hour for the next three hours. The Time Slider will display the data according to
    the time it was collected.

2   Open the properties of the Site1_Time layer and go to the Time tab.

    In the Time dialog box, you will identify the field that contains the time information, its
    format, and the interval value.

3   Click "Enable time on this layer." Set Time Field to HourMeasur and Field Format to
    YYYY/MM/DD hh:mm:ss. Next, click Calculate to set the layer time extent (duration).
    Set Time Step Interval to 1.00 and verify that the units are set to Hours. Then select
    the "Display data cumulatively" check box. Click OK to close the dialog box.

6-1

6-2

6-3

With the duration and interval set for the layer, you can use the Time Slider window to view
the data.

4 Click the Time Slider tool on the Tools toolbar.

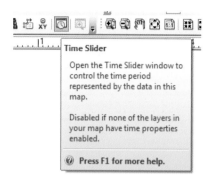

5 In the Time Slider window, click the Play button ▶ on the right.

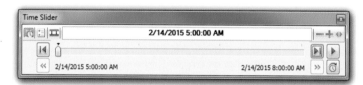

The data is shown in a time sequence. If the images go by too quickly, open the Time Slider options and set the speed slower. This time series is great for a visual representation, but it does not perform the analysis on building footprints that you did earlier in this tutorial. However, the flashy display is eye-catching. You can move the slider manually to show a specific time or let the series play.

All of this was strictly visual analysis, using common ArcMap tools and ALOHA freeware. The value is in knowing how to string together these tools to achieve a different type of analysis, which is to show how something changes over time.

6 Save your map document as **Tutorial 6-1.mxd** to the \GIST2\MyExercises folder. If you are not continuing to the exercise, exit ArcMap.

Taking it further

This tutorial uses some existing plume models for the match factory. The process of creating the plumes is a little long to include here, but you may be interested in learning how to create your own plume models. The tutorial uses the ALOHA plume-generating software, along with a custom script for importing the plume into ArcMap. These tools are free and can be downloaded from the National Oceanic and Atmospheric Administration website. To find them, go to http://response.restoration.noaa.gov and search for ArcTools.

Exercise 6-1

The tutorial showed the sequencing of a chemical plume and the process of selecting features inside the plume. In this exercise, you will repeat the process using another set of plume models.

- Continue using the map document you created in this tutorial, or open Tutorial 6-1.mxd from the \GIST2\Maps folder.
- Move to the Site 2 Bookmark.
- Turn off the Site 1 Chemical Plume Group.
- Turn on the Site 2 Chemical Plume Group.
- Symbolize the layers as you did in the tutorial.
- Select the buildings that fall within each of the plumes.
- Calculate the number of buildings, and their use, that fall within each plume.
- Add an informative text box; and change the titles, colors, and legend to make a visually pleasing map.
- Create a Time Slider series with the Site2_Time.lyr file.
- Save the results as **Exercise 6-1.mxd** to the \GIST2\MyExercises folder.

What to turn in

If you are working in a classroom setting with an instructor, you may be required to submit the maps you created in tutorial 6-1 and demonstrate the Time Slider series for the tutorial and exercise.

Turn in a printed map or screen capture of the following:

      Tutorial 6-1.mxd
      Exercise 6-1.mxd

6-1
6-2
6-3

Tutorial 6-1 review

The tutorial shows the change over time of a chemical-laced cloud. This type of analysis uses duration, number of measurements, and measurement interval. The duration is three hours, three plume prediction datasets are created, and the interval is one hour. Plumes are used for a predictive map that takes an event and predicts what will happen afterward.

Rather than do a time series of maps, you elected to overlay all the datasets and use a transparency to control how the data is displayed. For the Fire Department, the single display map works better for distribution, whether on paper or a computer screen.

You also saw that once the datasets are created, you can use them for further analysis. They can be used to select other features or to perform other types of overlay functions. With the included Time field, they can be used for a time series analysis using the Time Slider tool.

Study questions

1. How might you determine an interval value for a project?

2. What choices do you have for displaying the results of a time series analysis?

3. How might one of the previous tutorials be designed as a time series analysis?

Other real-world examples

The city Public Works Department may map the flood level over time to predict street closing or evacuations.

The US Census Bureau might use historical data to show population growth. This analysis can then be used to plan transportation systems or distribute funding for various projects.

Soil patterns may be analyzed over time to determine erosion areas or the spreading of a river delta.

Tutorial 6-2

Mapping change in location and magnitude

Data that changes in location may also have an associated value or attribute that changes over time. Through symbology, you can map both the change in physical location as well as show how the associated attribute values change.

Learning objectives

- *Symbolize values*
- *Understand variations in attribute values*
- *Create time series maps*

Preparation

- *Read pages 165–67 in* The Esri Guide to GIS Analysis, *volume 1.*

Introduction

When values are associated with data that changes location, you must see if the values are also changing. If so, the data is changing in both location and magnitude. You can symbolize the change in location using a linear path, or an expanding polygon as done in tutorial 6-1. The change in magnitude can be shown by changing the symbol's color or size.

Using a line will not only be useful for showing the current track, but may also be used as a predictive tool. It will give the viewer an idea of how quickly the data is changing. If the line segments between locations are short, it shows that the location is changing slowly. Conversely, if the distance between locations is long, it shows that the data values are changing rapidly, considering that the study interval remains constant.

Scenario Last April, a tornado touched down in the northern part of the city. Through the local weather-gathering services, you were able to get the locations of the actual touchdowns, the tornado's magnitude, and the wind speeds as it moved through town. You want to map both the change in location as well as the changes in the tornado's strength and wind speed.

Data The tornado data was gathered through local observations. It includes the magnitude on the Fujita Tornado Damage Scale for tornado intensity, on a scale of 1 to 5, and the wind speed in miles per hour.

6-1
6-2
6-3

The building footprint data is derived from aerial photographs, and each building includes a field named Layer that represents the building's use. The codes are as follows:

- 1 = Single Family
- 2 = Multi-Family
- 3 = Commercial
- 4 = Industrial
- 5 = Government
- 6 = Utilities
- 7 = Schools
- 8 = Churches

Road and lot boundary layers are included for background interest.

Set graduated symbols

1   In ArcMap, open Tutorial 6-2.mxd.

On April 7, a tornado ripped through town, tearing the roofs off many structures and leaving a general path of destruction. The city council needs a map of the damage path to secure relief funding. It must show the tornado as a dot proportional to its magnitude; the path as a linear symbol, with a variable width representing wind speed; and the damage buffer, which was

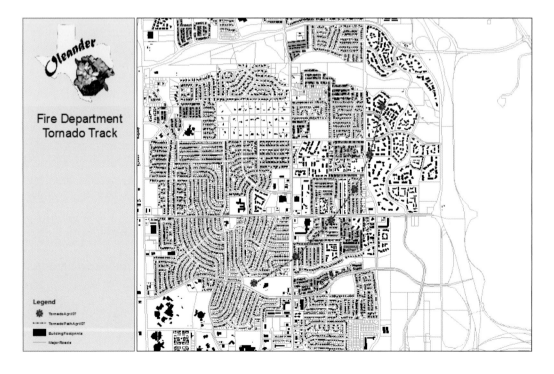

First, you will set the symbology for the point layer of tornado sightings. This symbology will indicate how strong the tornado was at each reading.

2   Open the properties of the TornadoApril07 layer. Set the symbology type to Quantities > Graduated symbols with the Fujita field for the value. Select an appropriate color and dot size range, as well as the number of classes in the classification. Click OK.

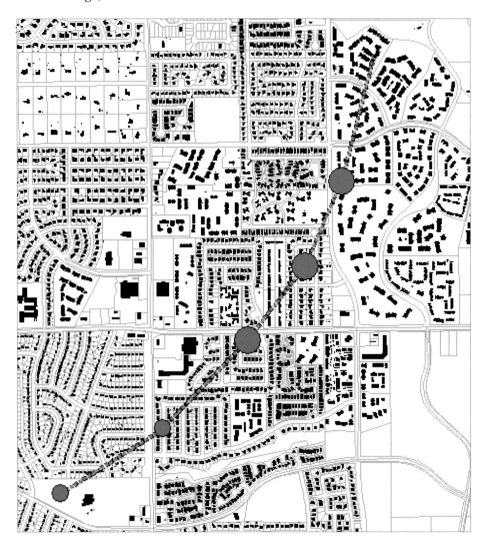

Now you will symbolize the path of the tornado, showing a depiction of measured wind speed.

6-1
6-2
6-3

3 Open the properties of the TornadoPathApril07 layer. Change the symbology type to Quantities > Graduated symbols with the Value field of WindSpeed. Select an appropriate color and line thickness range, as well as the number of ranges to display.

Two things can be read into this visual display of the storm data. First, it is easy to see how the tornado increased and decreased in intensity over time. Second, it is possible to see the increase and decrease in wind speed over the same time period.

Buffer the tornado's path

Next, you will add a buffer around the tornado's path that represents the damage area. Measurements were made, and the distance was stored in an attribute named BufferDist for your use in this part of the analysis.

1   Search for and open the Buffer tool. Set the following parameters:
      Input Features: TornadoPathApril07
      Output Feature Class: **TornadoPath_Buffer** in MyData.gdb
      Distance Field: BufferDist
      Dissolve Type: ALL
      When your dialog box matches the graphic, click OK.

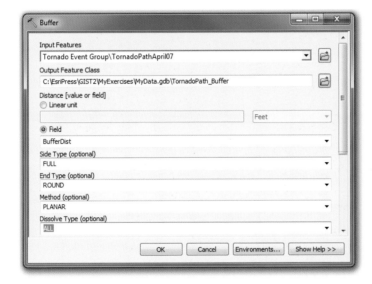

6-1
6-2
6-3

2   Open the properties of TornadoPath_Buffer. Set the color to Mango and the
transparency to 50%.

Select affected buildings

You will use this buffer area to select the affected structures. Use the following selection
statement to fill out the Select By Location tool:

*I want to select the features from the layer BuildingFootprints that intersect the features of the
layer TornadoPath_Buffer.*

Remember that if the BuildingFootprints layer does not appear in the selection dialog box,
it has not been made selectable.

1   Start the Select By Location tool and fill in the parameters suggested in the preceding statement. When your dialog box matches the graphic, click OK.

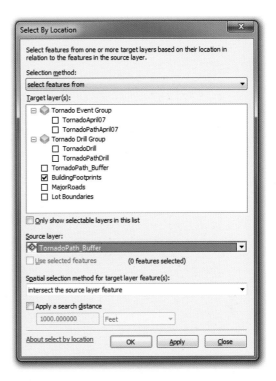

Each step in the process adds more valuable information. Now it is possible to see the visual display of the storm's intensity and the area of damage the storm caused.

2   Open the attribute table of the BuildingFootprints layer and build a summary process on the UseCode field of the selected buildings. Save the table as **BuildingTotal** to \GIST2\MyExercises\MyData.gdb.

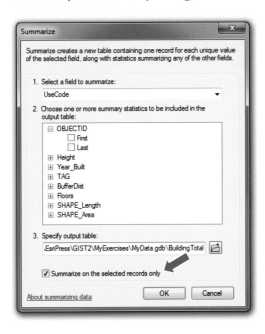

6-1
6-2
6-3

3 Clear the selection, and then add a text box to the title block to show the numbers of buildings damaged.

This map will now give the city council an idea of how many structures were damaged, as well as a visual representation of the wind speeds, tornado intensity, and ground speed of the tornado.

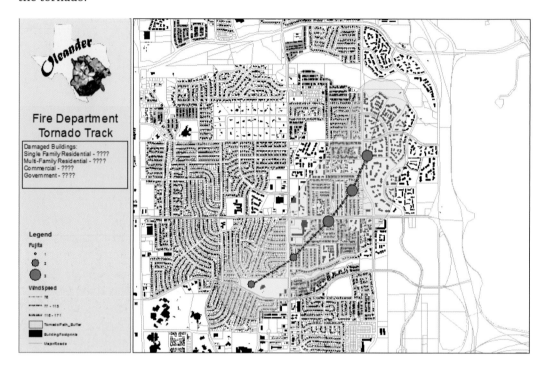

4 Save your map document as **Tutorial 6-2.mxd** to the \GIST2\MyExercises folder. If you are not continuing to the exercise, exit ArcMap.

Exercise 6-2

The tutorial showed how you can display numerous aspects of a tornado using different symbology and be able to read damage area, ground speed, and wind speed of the tornado.

In this exercise, you will repeat the process using data that the fire chief created for a drill scenario. You will practice creating the map, and the firefighters will evaluate the damage, evacuation plan, and other aspects of their planned response.

- Continue using the map document you created in Tutorial 6-2, or open Tutorial 6-2. mxd from the \GIST2\Maps folder.
- Move to the Drill Area bookmark.
- Turn off the Tornado Event Group layers.
- Turn on the Tornado Drill Group layers.
- Symbolize the tornado touchdown locations and path as shown in the tutorial.
- Buffer the tornado path using the BufferDist field.
- Select the buildings that fall within the damage area.
- Calculate the numbers of buildings in each use category that fall within the damage area.
- Add an informative text box; and change the titles, colors, and legend to make a visually pleasing map.
- Save the results as **Exercise 6-2.mxd** to the \GIST2\MyExercises folder.

What to turn in

If you are working in a classroom setting with an instructor, you may be required to submit the maps you created in tutorial 6-2.

Turn in a printed map or screen capture of the following:

Tutorial 6-2.mxd
Exercise 6-2.mxd

6-1
6-2
6-3

Tutorial 6-2 review

The observations of the tornado event were logged, along with the estimated intensity on the Fujita scale. This dataset allowed you to do a graduated-symbol depiction of the storm intensity. You also worked with a linear event, the wind speed.

By showing these attributes on the line, you can use a graduated symbol for the lines as well. The combination of thick lines and large dots highlights the most intense periods of the storm.

The data can also be used for an overlay process. You buffered the lines by a distance based on the recorded damage path, and used this buffer to select buildings. This buffer can also be overlaid on the parcel values for a quick damage estimate.

So through a combination of symbology on the lines and points, you can effectively communicate the changes in both location and magnitude.

Study questions

1. How can you calculate the speed of an event based on the type of data you are using?

2. How can magnitudes related to location data, both point and linear, be symbolized?

3. What technology might be used to collect point or line data that demonstrates change in location?

Other real-world examples

The weather tracking stations might track hurricanes by noting the longitude-latitude and intensity of each reading. This dataset can be used for predictions before the fact or damage assessment after the fact.

The city Engineering Department might move a traffic-counting device along a busy street and take readings, and then use the data to model traffic patterns at intersections.

A wildlife agency might track an animal herd, noting both the location and population through a migration pattern. A study can show how the population changes as the animals move.

Tutorial 6-3

Mapping percentage change in value

Data summarized by area, such as census counts or property values, changes over time. The change can be shown as simple values in a map series or as a percentage change for the study interval.

Learning objectives

- *Understand database attributes*
- *Map percentage change*
- *Perform time analysis*

Preparation

- *Read pages 168–73 in* The Esri Guide to GIS Analysis, *volume 1.*

Introduction

You may have worked with data that is collected for a single event and summarized for a given area. This data may be the number of permits issued in a subdivision, the value of a piece of property, or the count of tigers in a wildlife reserve. For the one event, the numbers tell an important story.

When another round of data collection occurs, things get interesting. Instead of showing a single-count event, you can now show how the values change over time: the change in the number of permits issued, the percentage increase in property value, or the change in tiger density.

The difference in values can be shown simply as a change in the number, as a percentage, or included in another calculation such as density. To get the change in the count, subtract the old value from the new value. This number can be positive or negative; and might be symbolized with the negative values in red, the positive values in green, and the zero or no change values in white.

Seeing the change for something such as property value is hard to comprehend over a large area, so it may be useful to show the change as a percentage. To calculate the percentage, subtract the old value from the new value, divide by the old value, and multiply by 100.

6-1
6-2
6-3

Finally, there are other means of showing the change, such as a change in density. This change is simply the difference of the values divided by the areas they represent.

After the change values are calculated, you can use standard symbology techniques to display the results.

Scenario The city council is concerned with property values in the city. It recently approved a new homeowner's assistance program that will help owners get grants to remodel their homes, thus increasing the values. It has also instituted a streamlined permitting process so that homeowners can get the bulk of the work done quickly. You have been asked to make a map to highlight the target areas: single-family residential property that has decreased in value more than 3 percent since 2012.

Data The parcel data includes a field for the appraised values in 2012 (TaxVal12) and a field for the appraised values in 2014 (TaxVal14). You will use these values to calculate a percentage change.

Road and lot boundary layers are included for background interest.

Calculate percentage change in property values

1 In ArcMap, open Tutorial 6-3.mxd.

This map shows the parcels of Oleander with the symbology set to single symbol. You will add a field to the database, calculate the percentage change in appraised value, and resymbolize the data.

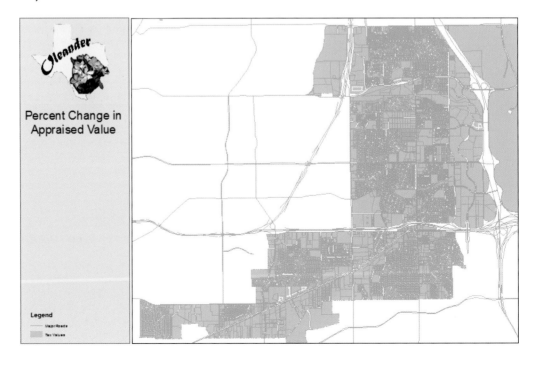

The program for funding passed by the city council is for residential property only, so step one will be to apply a definition query to the data to hide all the nonresidential property. You can do this query simply by using the field DU, which represents dwelling units and has been used in earlier tutorials. Only parcels that contain one dwelling unit qualify for this assistance.

2   Open the properties of the Tax Values layer and go to the Definition Query tab. Create the query DU = 1 and close the layer properties.

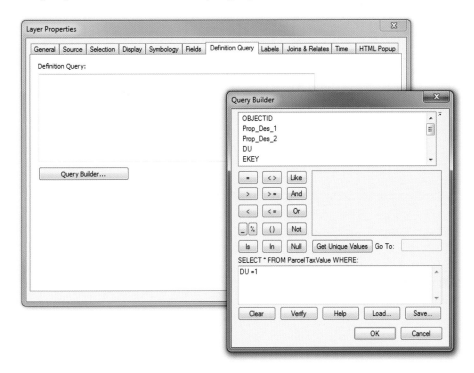

The next step will be to calculate the percentage change in tax value from 2012 to 2014. You will need a field to store the data, so you will add one to the attribute table. It will store decimal values, so its type is floating point.

6-1
6-2
6-3

3 Open the attribute table of the Tax Values layer and click Table Options > Add Field. Name the new field **PercentChange14** and set the field type to Float.

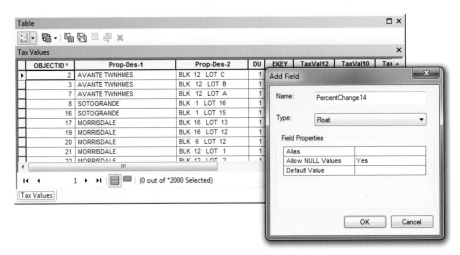

4 Verify that the new field has been added to the table, and then close or dock the attribute table.

A potential problem exists in the calculation for percentage difference. The 2014 data contains parcels that have recently been platted, but not yet added to the county's tax rolls and so are valued at zero. You might also find records in the 2012 data that have been taken off the tax rolls or combined on a common ownership account. Thus, you will also find values in the 2012 data with a value of 0.

You will add an expression to the definition query to remove these values from both datasets so that the calculation goes smoothly. If this precaution is not taken, you will likely get a "can't divide by zero" error. Will this new query be connected to the existing query using an AND or an OR? Write your answer before continuing.

5 Open the properties of the Tax Values layer, go to the Definition Query tab, and modify the query to exclude 0 values for TaxVal14 and TaxVal12 by adding the following: AND TaxVal14 <> 0 AND TaxVal12 <> 0. When your dialog box matches the graphic, click OK and then OK again to close the properties.

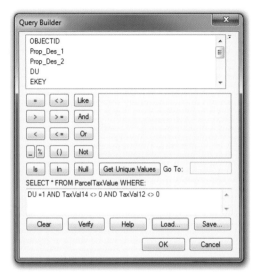

The results of this definition query will give you good values to use for the calculation. Several parcels are not included in this study, but they represent the parcels that have changed ownership in the past two years and do not qualify anyway. It is time to perform the calculation. The formula for percentage change in value, (New value – Old value) /Old value * 100, is as follows:

([TaxVal14] – [TaxVal12])/[TaxVal12] * 100.

6-1
6-2
6-3

6 Open the attribute table of the Tax Values layer. Right-click the new PercentChange14 field, click Field Calculator, and build the expression shown in step 5. (Read and dismiss the warning if you get one.) When your dialog box matches the graphic, click OK.

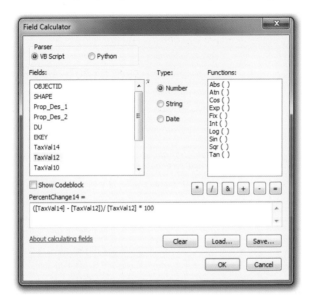

Notice that some values are 0, which means that their value did not change. There are some negative values and some positive values. Now the new values can be symbolized.

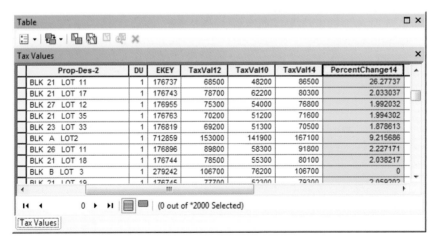

Note: once again, ArcMap gets a sample of features on which to base the range of feature values. The default maximum may be reached. You can either ignore this error or change the sample size value by clicking the Sampling button on the Classification dialog box and increasing the maximum.

Symbolize the results

The PercentChange14 field now holds a percentage change value that can be used to symbolize the results. A graduated-colors classification will make a good presentation. You may need to investigate how many classes to use. It is also a good idea to use a color scheme that shows positive results in green and negative results in red.

1 Close the attribute table. Open the properties of the Tax Values layer and go to the Symbology tab. Change the Show method to Quantities > Graduated colors and set the Value field to PercentChange14. Use the Color Ramp drop-down list to select a range from green to red.

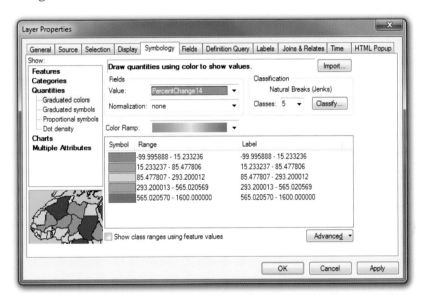

This setup looks good, but the ranges must be set manually to get the segregation needed. You want to show houses that had no change in yellow, houses with a 0 percent to 3 percent change in light green, and houses with a greater than 3 percent change in dark green. At the other end of the scale, show a 0 to –3 percent change in orange and a greater than –3 percent change in red.

6-1
6-2
6-3

2 Click the Classify button and set Classification Method to Manual. Change Break Values to -3.001, -0.001, 0, 3, and 1600. When your dialog box matches the graphic, click OK.

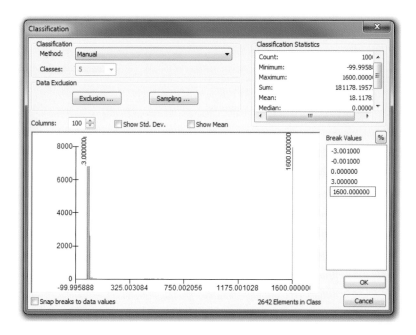

3 Change the labels to better describe the values, as shown in the graphic. Right-click the top symbol and click Properties for All Symbols. Set Outline Width to 0. Click OK and then OK again.

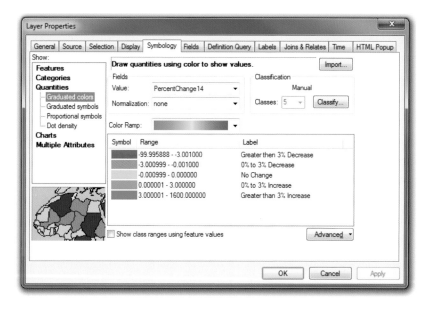

4   Turn off the Lot Boundaries layer momentarily and look for areas that have the highest negative change. Zoom to such an area. Turn the Lot Boundaries layer back on. The graphic shows the area identified by the bookmark Target Zone. You may have chosen a different area.

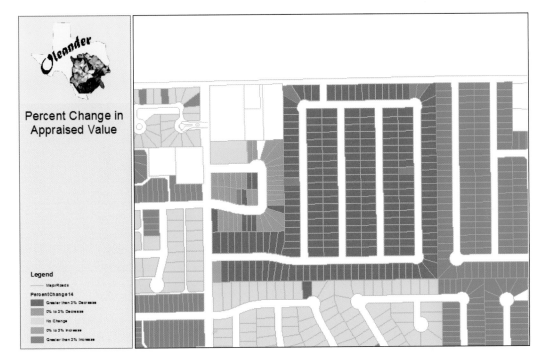

This resulting map will certainly give the city council an idea of where to concentrate their efforts to help stabilize the tax base. As it turns out, these houses are duplexes and might qualify for a federally funded project with matching money from the city. Using this information, the city council can leverage local tax dollars to do more good for the community.

5   Save your map document as **Tutorial 6-3.mxd** to the \GIST2\MyExercises folder. If you are not continuing to the exercise, exit ArcMap.

6-1
6-2
6-3

Exercise 6-3

The tutorial showed how to calculate a change in a value from the attribute table rather than a change in location.

In this exercise, you will prepare a map for the city council that shows the percentage change in tax value from 2010 to 2012. Use the fields TaxVal10 and TaxVal12.

- Continue using the map document you created in Tutorial 6-3, or open Tutorial 6-3.mxd from the \GIST2\Maps folder.
- Set the definition query for parcels with only one dwelling unit (DU).
- Add to the definition query a statement to exclude zero values.
- Add a field to the attribute table to contain the calculation.
- Calculate the percentage change.
- Set the classification manually to show no change, positive and negative changes of up to 3 percent, and positive and negative changes exceeding 3 percent.
- Zoom to an area that is a likely candidate for the remodeling program.
- Change the titles, colors, and legend to make a visually pleasing map.
- Save the results as **Exercise 6-3.mxd** to the \GIST2\MyExercises folder.

What to turn in

If you are working in a classroom setting with an instructor, you may be required to submit the maps you created in tutorial 6-3.

Turn in a printed map or screen capture of the following:

Tutorial 6-3.mxd
Exercise 6-3.mxd

Tutorial 6-3 review

Changes you mapped in tutorials 6-1 and 6-2 involved the features moving, as well as an attribute changing. There is an easy one-to-one relationship between the location change and the value change. In this tutorial, the feature did not move, but an attribute value changed. To show this change, you need a new attribute field that represents the change. If enough data is collected, you can create a time slider display as well.

This new value can represent the numeric change only, or it can be presented as a percentage, as you did in the tutorial. The change may also represent some characteristic of the parcel, such as whether it has been platted in the past year. Instead of calculating a value, you can assign a new code to these parcels and use the code to set its symbology.

Study questions

1. What role does time play in the tax data?

2. How might both the 2014 percentage change from the tutorial and the 2012 percentage change from the exercise be shown together?

3. Can you explain the unusually high percentage changes in tax value, such as a positive 1,600 percent increase?

Other real-world examples

The public library might want to map the change in the literacy rate for census tracts. Officials might look for areas with low literacy and initiate continuing education classes for reading. The effect of the reading campaign can be measured by showing new literacy levels over time.

Census data is all about change from one dataset to another, with the coverage areas remaining the same. But even the Census Bureau deals with situations in which the census blocks have zero values. These census blocks include areas such as industrial parks or airports, where there is no residential population. The inclusion of zeros in the database can create problems calculating a change percentage.

The water district might track the stored water in a reservoir and show the percentage change on a monthly basis. The polygons that represent the lakeshore will not change over time, but the measured water level will. A map series can show the water level measurement each month. The viewer can thus understand how the levels have changed over the study period.

6-1
6-2
6-3

7

Measuring geographic distribution

Many of the analysis techniques covered so far deal with the relationships of features to other features, but there is another realm of analysis that deals with the characteristics of features within a single dataset. Measuring the distribution of your data—that is, the distances between features within a dataset—will begin to help you understand the possibility that your data may have patterns or clustering. Measuring the compactness of the data and applying statistical methods to the measurements can reveal otherwise unseen characteristics in the data.

Tutorial 7-1

Calculating centers

A distribution of features in a dataset has many characteristics that are helpful for analysis. One of these characteristics is the calculated center of the distribution. Centers can point out trends in movement, the best location to service all the features, or the feature that is the shortest distance from other features.

Learning objectives

- *Identify the central feature*
- *Determine the median center*
- *Determine the mean center*

Preparation

- *Read pages 1–30 in* The Esri Guide to GIS Analysis, *volume 2.*

Introduction

Sets of features that are spread over an area are said to have a *geographic distribution*. Some sets of features might have a regular distribution, such as trees planted along a grid in an orchard. Other distributions might be seemingly random, such as locations of crime incidents. The goal of the spatial statistics tools in ArcGIS is to help identify and quantify spatial relationships and patterns. The results can then be used to compare datasets or track changes over time. The tools also help determine the statistical probability that the patterns and relationships really exist versus being random occurrences. Fortunately, the spatial statistics tools in ArcMap make this type of analysis much simpler. Regular GIS users do not need to be doctorate-level statisticians to use the tools. You only need to understand how the tools function and when to apply them to get valid results.

The first measurement of geographic distribution is finding the center. Three types of centers are calculated for data: the mean center, median center, and central feature. The formulas for calculating these values can be complex, and are illustrated in *The Esri Guide to GIS Analysis*, volume 2, on page 33.

The first type of center is the *mean center*. The mean center is calculated by finding the average of the x-coordinate values of all the features, and then finding the average of all the y-coordinate values. The resulting x,y coordinate pair is the mean center. Finding the mean center can be useful in determining the movement of a set of data over time, such as the migration of an animal herd.

The next type of center is the *median center*, also called the *center of minimum distance*. This function finds the x,y coordinate pair that is the closest to all the rest of the features. This function has a built-in geographic weighting, which does not occur with the mean center. Under this weighting, the median center gravitates toward areas with a lot of features. This type of analysis might be used to find a location for a new facility that is closest to all the other facilities. The median center tool is not demonstrated in this tutorial, but more information is available about its use in *The Esri Guide to GIS Analysis*, volume 2.

The third type of center is the *central feature*. This function identifies the single feature that has the lowest total distance to all the other features. Note that the central feature must occur at a location that has a feature, unlike the median center, which can occur at any x,y coordinate location. The Central Feature tool also has a built-in geographic weighting, in which groupings of features affect which feature is selected. For instance, this tool can be used to find the school campus that is most accessible to all the others for a districtwide meeting.

Scenario   The Fort Worth Fire Department in Tarrant County, Texas, has done some award-winning work with geospatial statistics. The fire chief, who is impressed with your previous maps, has agreed to allow you to work with the Fort Worth staff for this year's analysis and investigate the use of the spatial statistics tools. Not only will you be able to learn their techniques, but you can interject your own ideas and perhaps produce a better product.

They want to look at their calls for service data and fire station location data and make some determinations for future actions. Using the spatial statistics tools, they want to learn whether their past responses support their plans for current and future stations.

Data   The first dataset is the locations of the calls for service for February 2015. Each incident is located by both the address and a coordinate pair for the incident in the Texas State Plane, North Central projection. These locations will have a built-in geographic weighting because areas with a lot of calls for service tend to pull the centers in that direction.

The second set of data is the locations of the existing Fort Worth fire stations. This data includes the locations of proposed stations. You might be able to look at some of the proposed stations and determine whether the new locations would be better or worse for the recorded incidents.

Next are the alarm territories (AT). These territories are the primary response zones for each station. The Station field contains the number designation for each station. Several of these alarm territories are grouped to form a battalion zone, which is monitored by a battalion chief. Any major incident that occurs within an alarm territory will prompt the dispatching of the battalion chief for that battalion zone. The Battalion field in the attribute table contains the battalion zone number.

The fourth set of data is the street network data to give the map some reference. It comes from the Data and Maps for ArcGIS data.

7-1
7-2
7-3
7-4
7-5

Identify mean centers

1 In ArcMap, open Tutorial 7-1.mxd.

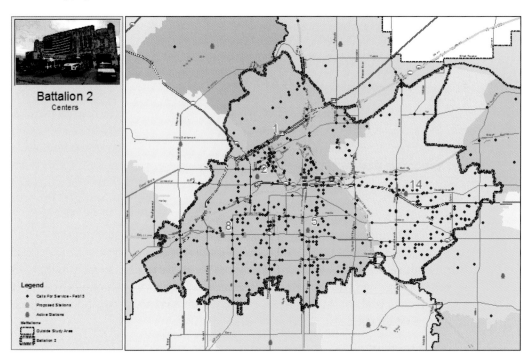

The map displays the Battalion 2 area, which is in the central part of Fort Worth. The existing stations are shown, as well as proposed stations. You will find the mean center of all the calls and the central feature of Battalion 2.

Battalion 2 contains several ATs, and you want to calculate the mean center of each AT. You will use the Mean Center tool to calculate it using the optional case parameter. You will use the Station field for the case value, causing the tool to calculate a mean center for each unique value in the Station field. Battalion 2 has five ATs, so you will get five mean centers. The dataset contains only the calls for service in Battalion 2 that originated from one of the Battalion 2 stations (Stations 1, 2, 5, 8, or 14).

You will run the Mean Center tool and compare the results with the station's location.

2 Use the Search window to locate and open the Mean Center tool.

3 In the Mean Center tool dialog box, set the parameters as follows:
- Input Feature Class: Calls For Service–Feb 15
- Output Feature Class: \GIST2\MyExercises\MyData.gdb\Bat_2_Mean_Centers
- Case Field: station

When your dialog box matches the graphic, click OK.

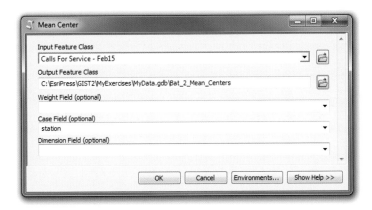

4　Turn off the Calls for Service–Feb 15 layer. Change the symbology of the Bat_2_Mean _Centers layer to a large yellow circle.

The result shows that most of the stations are located fairly close to their AT mean centers. Notice that the proposed new Station 8 is closer to the mean center than the existing station—this information might give the proposed location more credibility. It is important to note that when a fire unit from a station responds to a call outside its alarm territory, that location still affects the mean center.

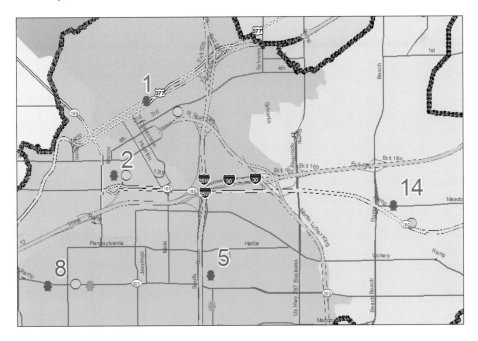

The Battalion 2 chief wants to relocate the battalion head office. The new location must be confirmed as the most centrally located of all the stations in Battalion 2 before the fire chief will approve the money. Battalion 2 includes Stations 1, 2, 5, 8, and 14. Next, you will select those stations and run the Central Feature tool.

Identify the central feature

Finding the central feature of Battalion 2 will first require that you select the active stations in that battalion. The tool will be run against the selected set.

1   On the main menu, click Selection > Select By Attributes. Build the following query to select the Battalion 2 stations from the Active Stations layer:

```
[BATALLION] = 2
```

2   Then click OK.

3   Locate and open the Central Feature tool.

An interesting option in the Central Feature tool is the Distance Method parameter. The Euclidean distance choice is the "as the crow flies" method, like the straight-line distances you worked with in tutorials 5-4 through 5-7. The Manhattan distance method moves from the source to the destination using vertical and horizontal lines that turn at right angles. Although this method does not equate the distance along a network analysis, as you did in tutorials 5-8 and 5-9, it is a better approximation in areas that use a regular street grid.

4   In the Central Feature dialog box, set the parameters as follows:
   • Input Feature Class: Active Stations
   • Output Feature Class: \GIST2\MyExercises\MyData.gdb\Bat_2_Central_Feature
   • Distance Method: MANHATTAN_DISTANCE

5 When your dialog box matches the graphic, click OK.

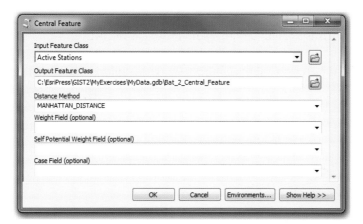

6 Set the symbology for the new layer to a large red star. Clear the selected features so that the symbology for all the layers displays correctly.

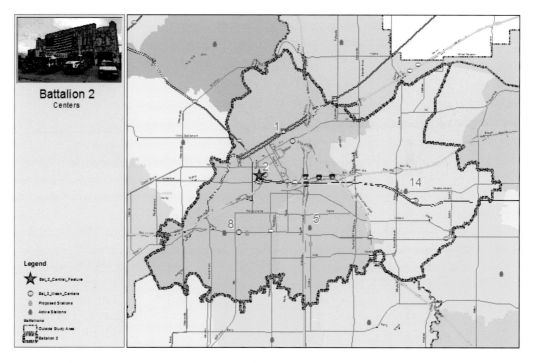

The central feature is Station 2. Setting up the office at Station 2 will ensure that the battalion chief is located in the station that is the most convenient to the other stations. Note that it does not mean that the chief will be any closer to expected calls for service.

7 Save your map document as **Tutorial 7-1.mxd** to the \GIST2\MyExercises folder. If you are not continuing to the exercise, exit ArcMap.

7-1
7-2
7-3
7-4
7-5

Exercise 7-1

The tutorial showed how to calculate a mean center and select a central feature.

In this exercise, you will repeat the process using different datasets. The Oleander city library has provided data about the patrons, and you want to locate three branch storefronts to serve them more conveniently. People can go to the storefront locations to use the computers, participate in certain library programs, and reserve books that can be collected later.

Part 1: find the mean center of each of the three districts.
- Open Tutorial 7-1E.mxd.
- Run the Mean Center tool for the Patron Locations layer. Set Case Field to A, which identifies the district.
- Symbolize the new layer appropriately.

Part 2: it is suggested that as part of the community outreach programs, the library should have on-site events at local apartment complexes. Using the Central Feature tool, find the best apartment complex in each district using the Apartment Complex polygons.
- Select District 1 in the Districts layer (use the AreaName field).
- Select Apartment Complexes in District 1.
- Find the central feature of Apartment Complexes.
- Repeat for Districts 2 and 3.
- Change the titles, colors, and legend to make a visually pleasing map.
- Save the result as **Exercise 7-1.mxd** to the \GIST2\MyExercises folder.

What to turn in

If you are working in a classroom setting with an instructor, you may be required to submit the maps you created in tutorial 7-1.

Turn in a printed map or screen capture of the following:
> **Tutorial 7-1.mxd**
> **Exercise 7-1.mxd**

Tutorial 7-1 review

Finding the centers of a set of features provided some interesting data for analysis. The mean center of the call for service data gives a good idea of the best place to locate a fire station. It would be interesting to calculate the mean center monthly and see how it moves throughout the year. With data over the time span of one year, it can be a useful predictive tool for winter calls versus summer calls.

Finding the central feature allowed the battalion chief to find a new office that might minimize his travel time. The total distance from each station to every other station is calculated, and the lowest total is chosen. Repeating this operation in the library exercise demonstrated that this function works on polygons as well.

It is important to remember that these two processes have a geographic weighting built in. Groupings of features tend to draw the center closer to them than areas with few or widely spaced features.

Study questions

1. Is it valuable to find the central feature of calls for service? Why or why not?

2. What other factors might be considered in locating a facility besides using the mean center of customers?

Other real-world examples

A police department may map crime locations by type and calculate the mean center to determine where to focus a task force.

A company may look for a new office location based on the mean center of its customers or employees.

A bank might schedule regional meetings at the branch calculated as the central feature to reduce overall travel time for employees.

7-1
7-2
7-3
7-4
7-5

Tutorial 7-2

Adding weights to centers

Considering all features equally when calculating centers may not give the most realistic results when other factors may be affecting the distribution of the values. Adding weights to these calculations will make the centers gravitate toward features with more importance in the analysis.

Learning objectives

- *Identify the weighted central feature*
- *Determine the weighted mean center*

Preparation

- *Read pages 30–38 in* The Esri Guide to GIS Analysis, *volume 2.*

Introduction

Some features in a geographically distributed set may carry more significance, or weight, than others. An option in the functions to calculate centers allows you to use these weights to draw the centers closer to these features. For instance, using the locations of businesses to calculate a mean center might put the location in one place. But factoring in the number of employees at each business might pull the mean center toward the businesses with more employees.

To add a weight to a center calculation, the features must have an attribute in their table that expresses the weight as a number. The attribute may be a count or a code, as long as it is a number that expresses the feature's importance.

Scenario   You are once again working with the Fort Worth Fire Department. It has provided a value for incident type that ranks its importance on a scale of 1 to 10. It seems that some of the calls logged in the database, such as false alarms and standby actions, were factored into the mean centers with equal weight. If the mean center method is used to evaluate the possible locations of fire stations, the chief wants to give higher-priority calls more significance in the calculations. The highest-priority calls involve the preservation of life, while medium-priority calls deal with the protection of property. The lowest-priority calls are false alarms and minor property destruction such as trash bin fires. You will recalculate the mean center using the weights provided.

Data   The Calls For Service–Feb 15 database has a field named FEE (Fire/EMS Evaluation). This field has values from 1 to 10, with 10 representing the most important calls.

In addition, the Active Stations layer has a field named NumEmployees that represents the number of employees at each station.

Identify weighted mean centers

1   In ArcMap, open Tutorial 7-2.mxd.

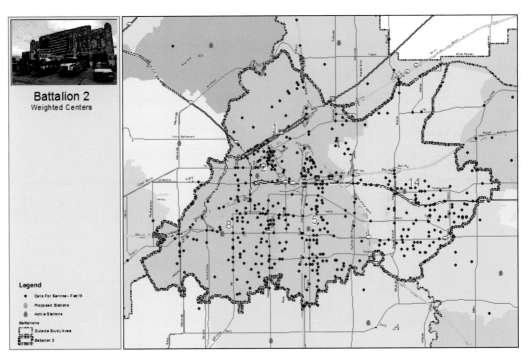

The map display looks just like tutorial 7-1. You will do the same process to calculate the centers, except this time use the FEE field as the weight for the mean center and the NumEmployees field as the weight for the central feature.

The data has already been restricted to only the calls for service from Stations 1, 2, 5, 8, and 14.

7-1
7-2
7-3
7-4
7-5

2 Open the attribute table of the Calls For Service–Feb 15 layer. Scroll through the records, noting the FEE and descript fields at the far right to get an idea of the incident types that get the highest weight. Close the table when you are finished.

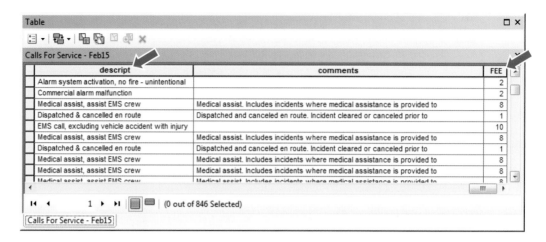

3 Locate and open the Mean Center tool and set the parameters as follows:

- Input Feature Class: Calls For Service–Feb 15
- Output Feature Class: \GIST2\MyExercises\MyData.gdb
  \Bat_2_Weighted_Mean_Centers
- Weight Field: FEE
- Case Field: station

4 When your dialog box matches the graphic, click OK.

5   Turn off the Calls For Service–Feb 15 layer and make the symbol for the new layer a large green circle. Add the layer Bat_2_Mean_Centers, which you created in tutorial 7-1, and symbolize the mean centers as a large yellow circle.

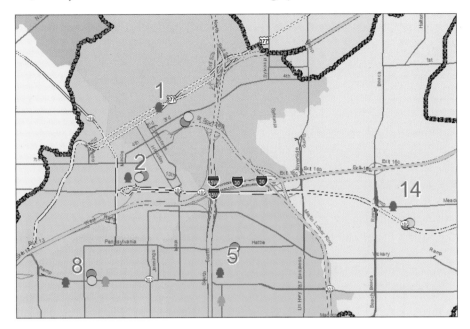

The weights have moved the mean center for each alarm territory slightly. In this case, the effect of the weights was minimal, but some weights may alter the results dramatically.

Next, you will try the Central Feature command using the NumEmployees field as a weight. The chief wants to be at the location that gives the best access to the firefighters. Remember that in tutorial 7-1, Station 2 was selected as the new home for the battalion chief.

Identify the weighted central feature

1   Use the Select By Attributes tool to select only the stations in Battalion 2 (Stations 1, 2, 5, 8, and 14).

2   Locate and open the Central Feature tool and set the parameters as follows:

- Input Feature Class: Active Stations
- Output Feature Class: \GIST2\MyExercises\MyData.gdb \Bat_2_Weighted_Central_Feature
- Distance Method: MANHATTAN_DISTANCE
- Weight Field: NumEmployees

7-1
7-2
7-3
7-4
7-5

3 When your dialog box matches the graphic, click OK.

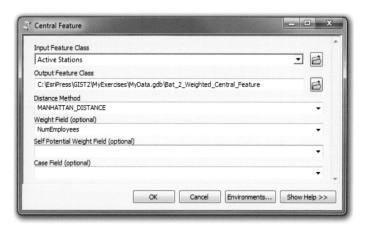

With some features having more significance in the calculation, or a larger weight value, you can expect to see a shift in the location of the central feature.

4 Clear the selected features and change the symbology of the new central feature layer to a large green star.

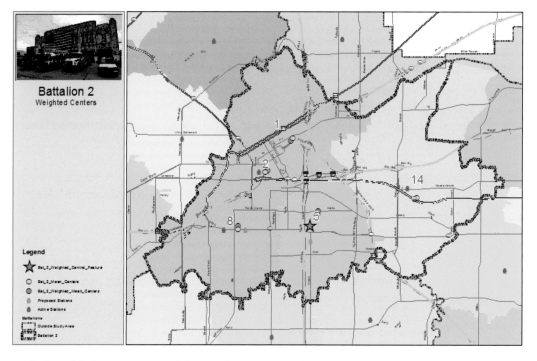

With the added importance of the employee count, the new battalion chief location has moved from Station 2 to Station 5.

5 Save your map document as **Tutorial 7-2.mxd** to the \GIST2\MyExercises folder. If you are not continuing to the exercise, exit ArcMap.

Exercise 7-2

The tutorial showed how to add an optional weight value, which gives some features more significance in the calculation to find the center.

In this exercise, you will repeat the library analysis from exercise 7-1, adding a weight value. The library database is summed for each address, so the layer Patron Locations contains a field that sums the number of visits to the library from each address. Some of the addresses are apartment complexes, so the number for these addresses gets high.

Part 1: find the weighted mean center for each of the three districts. The field A in the table for Patron Locations can be used as the case field, and the field Usage can be used as the weight field.
- Open Tutorial 7-2E.mxd.
- Run the Mean Center tool for the Patron Locations layer. Set Weight Field to Usage and Case Field to A.
- Symbolize the output appropriately.

Part 2: the Apartment Complex layer contains a field named DU, which represents the number of dwelling units at each site. Use the DU field as the weight field for the central location calculation so that it takes into account the number of units, and not only the geographic location.
- Select District 1 in the Districts layer.
- Select apartment complexes in District 1.
- Find the central feature of the apartment complexes, with the weight field set to DU.
- Repeat for Districts 2 and 3.
- Change the titles, colors, and legend to make a visually pleasing map.
- Save the results as **Exercise 7-2.mxd** to the \GIST2\MyExercises folder.

What to turn in

If you are working in a classroom setting with an instructor, you may be required to submit the maps you created in tutorial 7-2.

Turn in a printed map or screen capture of the following:
>   **Tutorial 7-2.mxd**
>   **Exercise 7-2.mxd**

Tutorial 7-2 review

Adding a weight to the mean center calculations allowed you to see how another characteristic of the features can affect the results. The features had a built-in geographic weight, but other attributes such as a value or count can further enhance the reality of the calculation.

The inclusion of a weight also affects the central feature calculation. By adding another characteristic to the calculation, the result is drawn toward the feature with the largest weight value.

Study questions

1. What types of values might be useful as a weight value? Which types might not be?

2. What is the purpose of adding a weight to the calculations?

Other real-world examples

A police department may weight crime locations by severity to assign a specific task force to the areas with the most dangerous crimes.

A company planning to open a new office may weight the calculation of a central location by the number of transactions attributed to each customer. The result will be the location closest to the most active customers, thus reducing overall customer travel time.

A bank looking for a site to hold regional meetings might weight the calculation of a central location by the number of employees at each branch to reduce the overall driving distance for all employees.

Tutorial 7-3

Calculating standard distance

Measuring the compactness of a distribution provides a single value that represents the dispersion of features around the center. This value is calculated by measuring the average distance of the features from the mean center and by how much the individual distances vary from the average distance. This value is called the *standard distance deviation*, or simply the *standard distance*.

Learning objectives

- *Calculate the standard distance*
- *Determine the compactness of distribution*
- *Work with weight values*

Preparation

- *Read pages 39–44 in* The Esri Guide to GIS Analysis, *volume 2.*

Introduction

Measuring the standard distance allows you to quantify the amount of dispersion in a set of features. It is calculated by determining the average distance each feature is from the mean center, and then determining a value for how much the distances deviate from the average distance. The results are shown on a map as circular rings of one, two, or three standard deviations. As with the calculation of mean centers, a natural geographic component of each feature is used to calculate standard distance.

Standard distance circles can also be given a weight, similar to the weighted centers. Besides the natural geographic components, an extra factor can be introduced to influence the results. If a weight is given, the results are measured around the weighted mean center.

It is important to note that a standard distance circle does not determine grouping or clustering, only that the features are occurring close by. A standard distance circle may include several groupings or clusters. These topics will be discussed later in chapter 8.

Scenario   You will continue the analysis of the Battalion 2 data for the Fort Worth Fire Department. The goal is to show the standard distance for each station. You want to calculate the standard distance circle both with and without the Weight value FEE, which will add more importance to some features than others. From the results, you will be able to determine which station's units are making more runs outside their normal alarm territory.

7-1
7-2
7-3
7-4
7-5

Data   the Calls For Service–Feb 15 database has values from 1 to 10, with 10 representing the most important calls.

The FEE field in The last set of data is the street network data to give the map some reference. It comes from the Data and Maps for ArcGIS data.

Determine standard distance

1   In ArcMap, open Tutorial 7-3.mxd.

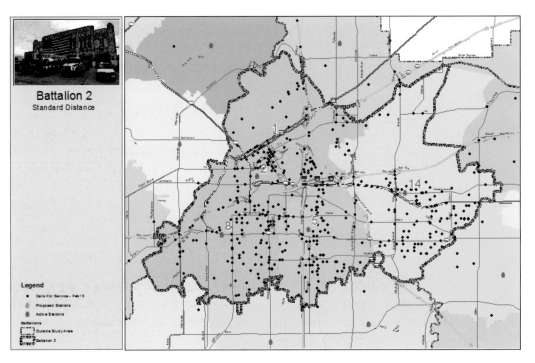

The five stations from Battalion 2 will be used for the standard distance calculations. You will compare across districts, which may pinpoint stations that are understaffed.

2   Locate and open the Standard Distance tool.

3   In the Standard Distance dialog box, set the parameters as follows:
- Input Feature Class: Calls For Service–Feb15
- Output Feature Class: \GIST2\MyExercises\MyData.gdb\Bat_2_Stnd_Dist
- Circle size: 1_STANDARD_DEVIATION
- Case Field: station

4   When your dialog box matches the graphic, click OK.

The results are five circles, one standard deviation from the mean center of each alarm territory.

5   Set the symbology to hollow with a thick blue outline.

You can see that Stations 2, 5, and 8 have a fairly compact standard distance circle. Their average travel time to incidents is mostly within the boundaries of their alarm response territory. Stations 1 and 14 have large standard distance circles, which shows that their average travel time is relatively long.

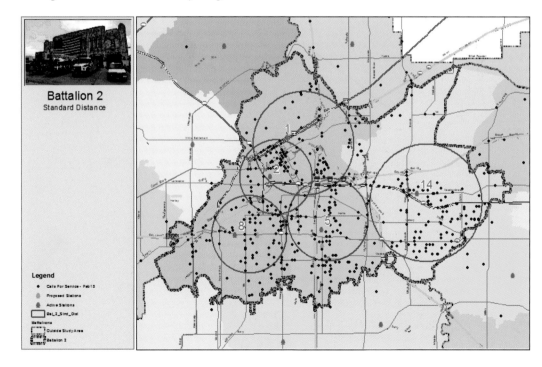

7-1
7-2
7-3
7-4
7-5

These circles do not take into account the relative importance of the incidents. You are more concerned with the average travel time to the most important calls. So if the more critical calls are adding to the size of the standard distance, you might want to consider adding another station to lower the average response time. Run the process again using the field FEE as the weight.

Determine weighted standard distance

1 Open the Standard Distance tool and set the parameters as follows, and then run the tool:

- Input Feature Class: Calls For Service–Feb 15
- Output Feature Class: \GIST2\MyExercises\MyData.gdb\ Bat_2_Weighted_Stnd_Dist
- Weight Field: FEE
- Case Field: station

2   Set the new layer to hollow with a thick yellow outline.

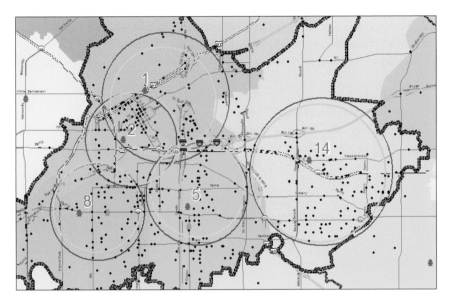

Compare the weighted and unweighted standard distance circles. Instances in which the weighted standard distance circle is smaller or more compact than the unweighted standard distance circle demonstrate that the inclusion of the priority code as a weight impacts the performance measurements of the stations. The analysis of Stations 5 and 8 already demonstrated compactness as unweighted standard distance circles and did not change much with the inclusion of a weight value. However, the analysis of Stations 1 and 14 shows a more compact standard distance circle when the weight factor is included. Overall, the compactness of the standard distance circles for all the stations is better when the incident priority code is added as a weight to the analysis.

3   Save your map document as **Tutorial 7-3.mxd** to the \GIST2\MyExercises folder. If you are not continuing to the exercise, exit ArcMap.

7-1
7-2
7-3
7-4
7-5

Exercise 7-3

The tutorial showed how to calculate both weighted and unweighted standard distance circles to assess the compactness of feature distribution.

In this exercise, you will repeat the process using the data from the library again. You will compute the standard distance for each district, and then repeat it using a weight. The weight field is Usage, which represents the number of library transactions attributed to each feature. Note that some usage values seem unusually high because they are summed for apartment complexes.

Part 1: find the standard distance for each of the three districts. The field A in the table for Patron Locations can be used as the case field.
- Open Tutorial 7-3E.mxd.
- Run the Standard Distance tool. Set Case Field to A.

Part 2: run the Standard Distance tool again with a weight. The field A is again used as the case field, and the weight field is Usage.
- Change the titles, colors, and legend to make a visually pleasing map.
- Save the results as **Exercise 7-3.mxd** to the \GIST2\MyExercises folder.

What to turn in

If you are working in a classroom setting with an instructor, you may be required to submit the maps you created in tutorial 7-3.

Turn in a printed map or screen capture of the following:
> **Tutorial 7-3.mxd**
> **Exercise 7-3.mxd**

Tutorial 7-3 review

You created both weighted and unweighted standard distance circles in this tutorial. It is apparent that many of the features used for standard distance fall within the circle, while other features fall outside the circle. The features outside affect the size of the circle, and thus the compactness of the distribution. With the weight field applied, the features outside the circle may not affect the calculations in a negative way if their weight value is low. In most cases, you saw that adding a weight reduces the size of the circle.

It is important to note that features that may belong to different geographic clusters still have a single circle to represent the distribution. Once again, the standard distance circles show compactness of the overall distribution, but do not show groupings or clusters.

Study questions

1. What does the Standard Distance tool measure?

2. What is the significance of the weight when calculating a standard distance?

3. What does the data distribution chart look like for data with a small standard distance? And with a large one?

Other real-world examples

A police department may measure the compactness of a set of crime data to determine a possible patrol area. A concentrated area can be easy to patrol with extra officers, while a dispersed area may not highlight a clear tactic.

A county health service may look at the standard distance for West Nile virus cases to determine a mosquito-treatment area. A small standard distance circle may highlight an area in which to focus treatment.

7-1
7-2
7-3
7-4
7-5

Tutorial 7-4

Calculating a standard deviational ellipse

Compactness is only one characteristic of distribution that may be of interest. Another one is directional trend. The standard deviation is calculated separately for the x- and y-axes, resulting in an *elongated circle*, or *ellipse*.

Learning objectives

- *Compute standard deviational ellipses*
- *Identify directional trends*
- *Work with weight values*

Preparation

- *Read pages 45–50 in* The Esri Guide to GIS Analysis, *volume 2.*

Introduction

The distribution of features may have a directional trend, making it hard to represent with a circle. By calculating the variances for both the x- and y-axes independently, you can show the directional trend. The result is shown with an ellipse, which orients the axes to match the orientation of the features.

The ellipse is referred to as the *standard deviational ellipse* because the method calculates the standard deviations of the x-coordinates and y-coordinates from the mean center. When the distribution is elongated along an axis, the ellipse reflects the orientation of that axis. Although visual analysis may give you an idea of the orientation of the features, the standard deviational ellipse will give you more confidence in your results because it is based on a statistical calculation.

With the addition of an orientation component, standard deviational ellipses can more accurately depict the distribution of data than the standard distance. Features that group along a path can create a large standard distance circle, but a thin, elongated standard deviational ellipse. For instance, the orientation of the ellipse may reveal a trend along a road or stream.

As with the standard distance, you can calculate the standard deviational ellipse in two ways. You can use either the locations of the features as a built-in geographic weight, or allow an attribute value to influence the locations. The latter method is the *weighted standard deviational ellipse*.

Scenario   You are going to revisit the Battalion 2 data for the Fort Worth Fire Department. The freeways going through the region may give more of a directional trend to the data that can be revealed using the standard deviational ellipse. You want to calculate the ellipse both with and without a weight based on the FEE attribute, which will add more importance to some features than others. From the results, you will be able to visualize any directional influence that the roads may have on the data.

Data   the Calls For Service–Feb 15 database has values from 1 to 10, with 10 representing the most important calls.

The FEE field in The map also includes street network data to give the map some reference. It comes from the Data and Maps for ArcGIS data.

Use the Directional Distribution tool

1   In ArcMap, open Tutorial 7-4.mxd.

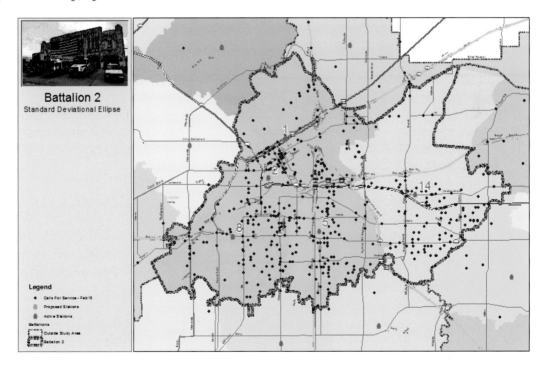

Once again you get the familiar Battalion 2 area. These five stations will be used for the standard deviational ellipse calculations. Comparing the ellipses across districts may give you an idea of any directional trends.

2   Locate and Open the Directional Distribution tool.

3   In the Directional Distribution dialog box, set the parameters as follows:
- Input Feature Class: Calls For Service–Feb 15
- Output Ellipse Feature Class: \GIST2\MyExercises\MyData.gdb\ Bat_2_Stnd_Dev_Ellipse
- Case Field: station

4   When your dialog box matches the graphic, click OK.

The standard deviational ellipse is calculated and added to the table of contents, with one ellipse for each occurrence of the case field. Note the shape and orientation of the ellipses, and match the directional trend to both the roads and the features.

5   Set the symbology of the Bat_2_Stnd_Dev_Ellipse layer to a hollow symbol with a thick blue outline.

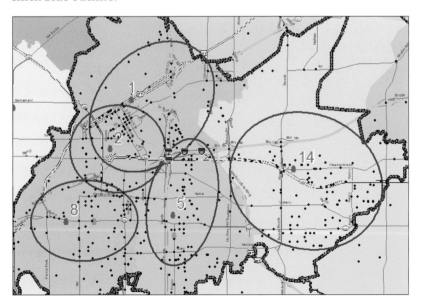

You can easily see that the calls from Station 5 have a definite north–south trend, following the freeway. Station 8 has an east–west trend, following another freeway. Station 2's orientation is affected by calls in a particular neighborhood, and not oriented to the roads; and Station 14's broad response area and even distribution keep its ellipse very close to being a circle.

Next, you will try this analysis again using the rankings in the FEE field as a weight.

Use the Directional Distribution tool with weights

1   Open the Directional Distribution tool again and set the parameters as follows:

- Input Feature Class: Calls For Service–Feb 15
- Output Ellipse Feature Class: \GIST2\MyExercises\MyData.gdb\ Bat_2_Weighted_Stnd_Dev_Ellipse
- Weight Field: FEE
- Case Field: station

2   When your dialog box matches the graphic, click OK.

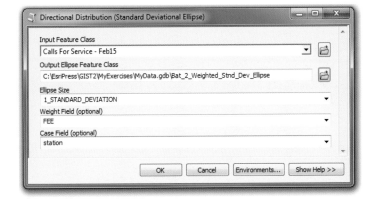

3 Set the symbol for the Bat_2_Weighted_Stnd_Dev_Ellipse layer to hollow with a thick yellow outline.

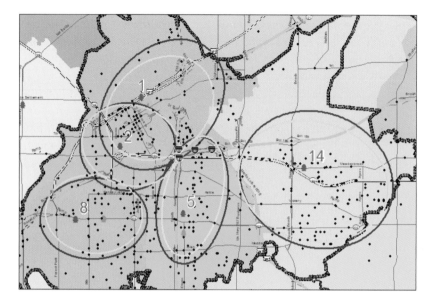

Notice that the orientation of Station 5's ellipse is narrower with the weights applied, and the distribution of Station 14 calls is less circular. This analysis is for only one month's worth of data, and you will want to look at the change in the ellipses over a longer period to determine the most optimal station placement.

4 Save your map document as **Tutorial 7-4.mxd** to the \GIST2\MyExercises folder. If you are not continuing to the exercise, exit ArcMap.

Exercise 7-4

The tutorial showed how to calculate both weighted and unweighted standard deviational ellipses. The results not only show compactness of the data, but also highlight directional trends.

In this exercise, you will repeat the process using the data from the library again. You will compute the standard deviational ellipse for each district, and then repeat it using a weight. The weight field is Usage, which represents the number of library transactions attributed to each feature. Note that some usage values seem unusually high because they are summed for apartment complexes.

Part 1: find the standard deviational ellipse for each of the three districts. The field A in the table for Patron Locations can be used as the case field.
- Open Tutorial 7-4E.mxd.
- Run the Directional Distribution tool. Set Case Field to A.

Part 2: run the Directional Distribution tool again with a weight. The field A in the table for Patron Locations is again used as the case field, and the weight field is Usage.

Part 3: add the weighted standard distance circles from exercise 7-3 and compare the results.
- Change the titles, colors, and legend to make a visually pleasing map.
- Save the results as **Exercise 7-4.mxd** to the \GIST2\MyExercises folder.

What to turn in

If you are working in a classroom setting with an instructor, you may be required to submit the maps you created in tutorial 7-4.

Turn in a printed map or screen capture of the following:
      Tutorial 7-4.mxd
      Exercise 7-4.mxd

7-1
7-2
7-3
7-4
7-5

Tutorial 7-4 review

You created both weighted and unweighted standard deviational ellipses. In addition to showing data compactness, like the standard distance circle, the ellipse can show a directional trend. The weights for the fire response data tended to draw the ellipse toward the freeways, where many traffic accidents may have a large effect.

If the ellipses remain more circular, it indicates that the data does not have a directional trend. The data shows that calls from Station 14 were spread all across the district. If many of the calls for service around the edges of the district have a large weight, the ellipse will not appear much different from the standard distance circle. A resulting ellipse that is more circular indicates that the data does not have much directional trend.

Study questions

1. What information do the ellipses provide that the circles do not?

2. What is the relationship between the weighted center and the weighted standard deviational ellipse?

3. What is the difference between the weighted standard deviational ellipse and the weighted standard circle?

Other real-world examples

A police department may look at the standard deviational ellipse to spot directional trends in crime, such as following a hiking trail or freeway.

Animal-tracking data may be mapped with a standard deviational ellipse to see directional trends in migration or grazing.

Tutorial 7-5

Calculating the linear directional mean

The directional trend of linear objects is calculated using a different methodology from points or polygons. The lines are analyzed to determine their angles, and then the mean angle is calculated. The result demonstrates a directional mean as an arrow.

Learning objectives

- *Identify directional trends of line features*
- *Compute the linear directional mean*
- *Work with case values*

Preparation

- *Read pages 51–61 in* The Esri Guide to GIS Analysis, *volume 2.*

Introduction

Calculating the linear directional mean can result in one of two outputs, depending on the input data. The outputs are the *mean direction* and the *mean orientation*. Each result is based on the average angle of the lines in the dataset, but mean direction takes into account the direction of movement.

When lines are created in ArcMap, they have a start point and an end point. The deviation of a line from the horizontal is measured as an angle from the start point to the end point. Then each of the lines is transposed so that their start points are coincident, and a mean angle is calculated. The mean direction is useful for data that represents features that move, such as hurricanes or wildlife. The results of the Linear Directional Mean tool are then displayed using an arrow to show direction.

If the line features represent stationary items, such as fault lines or freeway systems, they do not have a direction. Instead, the line created by the Linear Directional Mean tool shows only orientation. The line might be shown with a double-headed arrow.

7-1
7-2
7-3
7-4
7-5

Before using this tool, be sure to understand which result you will get. If you want the mean direction, make all the line features point in the correct direction regarding their start and end points. Also note that for multisegment lines, only the start and end points of the feature are used to calculate that feature's angle. Vertices are not used in the calculation. So an S-shaped feature is represented in the equation by a single line, from start point to end point.

Scenario   As part of its disaster planning, the Tarrant County Emergency Operations Center wants to analyze past tornado events in the county. After some research, archival data was assembled into a GIS format. You are asked to find the linear directional mean of each year's storms to better predict future occurrences.

Data   The layer Tornado has single point measurements of tornado touchdowns. The attribute table includes the year, latitude-longitude coordinate pair, and the tornado's measured intensity on the Fujita scale.

The layer Storm Track contains lines that connect the tornado touchdown points for each individual storm. Because the lines were digitized by connecting the first touchdown point of each tornado to its last, they will produce not only an angular trend, but also a directional trend.

The other layers are the census population data and road network for background interest.

Determine the linear directional mean of tornadoes

1   In ArcMap, open Tutorial 7-5.mxd.

The map shows the current population in purple, along with a street network. The storm tracks are the paths that each year's recorded tornadoes took after touchdown. You will calculate the linear directional mean to see what the directional trend was for each year's storms. This analysis will give the fire department an idea of where future tornadoes may go after touchdown.

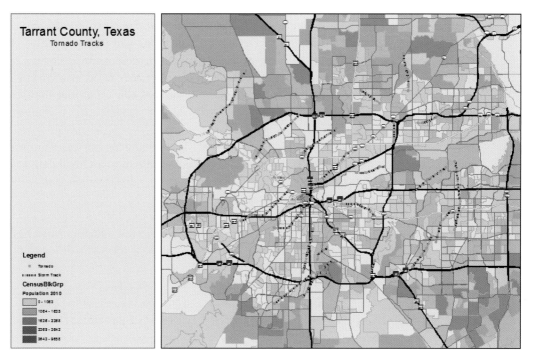

Tarrant County, Texas
Tornado Tracks

Legend
○   Tornado
▪▪▪▪ Storm Track
CensusBlkGrp
Population 2010
□ 0 - 1063
▨ 1064 - 1635
▨ 1636 - 2268
▨ 2269 - 3642
▨ 3643 - 9655

2 Locate and open the Linear Directional Mean tool.

3 In the Linear Directional Mean dialog box, set the parameters as follows:
  - Input Feature Class: Storm Track
  - Output Feature Class: \GIST2\MyExercises\MyData.gdb\Storm_Linear_Dir_Mean
  - Case Field: YEAR_

4 When your dialog box matches the graphic, click OK.

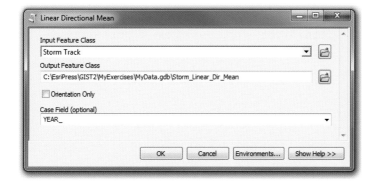

7-1
7-2
7-3
7-4
7-5

The results are lines that represent the directional means of the sets of storms for each year. To show direction, the symbology has been changed to an arrow.

5 Open the properties of the new Storm_Linear_Dir_Mean layer and go to the Symbology tab. Click the symbol to open the Symbol Selector, and set the symbol to Arrow at End, Mars Red, and a width of 5. (See if you can figure out how to make the arrowhead red as well.)

6 Next, go to the Labels tab of the properties and set Label Field to YEAR_, font size to 14, and color to red. Select the check box to turn the labels on and exit the properties dialog box.

The storms seem to move in the northeast direction. To see the trend for a specific year, select only the 1977 storms, for instance.

7 On the main menu, click Selection > Select By Attributes. Using the Storm Track layer, build the query YEAR_ = '1977'. When your dialog box matches the graphic, click OK.

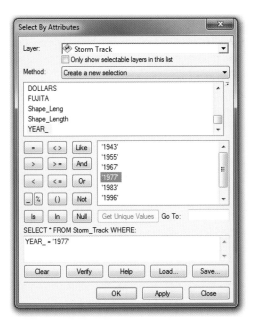

You can see the four storm tracks that are used to make the 1977 directional mean line. The directional mean was calculated, and a line drawn at the mean angle was added at the mean center of the selected storm tracks. The length of the directional mean line is the average length of the storm track lines.

7-1
7-2
7-3
7-4
7-5

YOUR TURN

Select the storm tracks for the other years to see their trends. Open the Select By Attributes dialog box and build similar queries for 1943, 1955, 1967, 1983, 1996, and 2003. Compare the lengths of the selected storm track lines with the length of the calculated linear directional mean for each year's set of tornado tracks.

8 Save your map document as **Tutorial 7-5.mxd** to the \GIST2\MyExercises folder. If you are not continuing to the exercise, exit ArcMap.

Exercise 7-5

The tutorial showed how to calculate the linear directional mean for the tornado tracks in Tarrant County. The results give you the directional trends of the storms, and the mean path length of each year's storms.

In this exercise, you will repeat the process using the data from Dallas County. The storm tracks are mapped. The YEAR_ field in the Storm Track Dallas attribute table contains the year that each storm occurred.

- Open Tutorial 7-5E.mxd.
- Run the Linear Directional Mean tool. Set Case Field to YEAR_.
- Compare the path lengths of each year's set of storms with the length of that year's linear directional mean.
- Classify the census data to show population.
- Change the titles, colors, and legend to make a visually pleasing map.
- Save the results as **Exercise 7-5.mxd** to the \GIST2\MyExercises folder.

What to turn in

If you are working in a classroom setting with an instructor, you may be required to submit the maps you created in tutorial 7-5.

Turn in a printed map or screen capture of the following:
> **Tutorial 7-5.mxd**
> **Exercise 7-5.mxd**

7-1
7-2
7-3
7-4
7-5

Tutorial 7-5 review

The Linear Directional Mean tool created single lines to represent the directional trend of the tornadoes in the study area. Because the storm tracks have a direction, and the lines are verified to be digitized in the correct order, the results can be symbolized with an arrow.

The length of the output line represents the mean length of the storm tracks. Keep in mind, though, that path length may not be indicative of the damage the storm caused.

Study questions

1. Which of these two datasets will produce the longest output vector: a lot of short storm tracks or a few long ones? Why?

2. If the fire department knows the directional mean of storms from the past 10 years, what planning might it be able to do for an approaching storm?

3. How does mean orientation differ from mean direction?

Other real-world examples

A police department can map the track from the point a car is stolen to the point of recovery. The results give a directional mean that can help predict the actions of an auto theft ring.

Wildlife biologists use animal-tracking data to map migration paths. The linear directional mean of these paths may give some insight into why certain paths are chosen or highlight the need for protected habitat.

Transportation studies often match traffic counts with the mean orientations of freeway systems. The results may show where there are too few roads to handle the traffic and may help plan the system's expansion.

8

Analyzing patterns

The statistical analysis tools used to study patterns in a dataset can answer a global question: What is the probability that the distribution of these features is occurring due to random chance? These tools do not create anything that can be symbolized on a map, although it may be helpful to capture some of the graphics that they create. The tools are important, however, in establishing a statistical foundation as well as a statistical confidence level for pattern analysis.

Tutorial 8-1

Using average nearest neighbor

To identify patterns when working with spatial statistics, various tools are used to measure a characteristic of the data that indicates whether the data is thought to be clustered or dispersed; or if, in fact, it occurs randomly. This measure, or *index*, is then tested to see if it can be believed statistically with some degree of confidence. The Average Nearest Neighbor tool calculates an index that reflects the average distance from a feature to all its neighbors compared with the average distance for a random distribution.

Learning objectives

- *Test for statistical significance*
- *Evaluate z-scores*
- *Understand random distribution*
- *Understand clustering versus dispersion*

Preparation

- *Read pages 63–79 and 88–96 in* The Esri Guide to GIS Analysis, *volume 2.*

Introduction

The Average Nearest Neighbor tool is used to determine if a set of features shows a statistically significant level of clustering or dispersion. It does this by measuring the distance from each feature to its single nearest neighbor, and then calculating the average distance of all the measurements. The tool next creates a hypothetical set of data with the same number of features but placed randomly within the study area. It then calculates an average distance for these features and compares these distances to the real data. A nearest neighbor index is returned, which expresses the ratio of the observed distance divided by the distance from the hypothetical data. If the number is less than one, the data is considered to be exhibiting clustering; if it is more than one, the data is exhibiting a trend toward dispersion.

The idea behind this tool is that things near each other are more alike than things that occur far apart, an axiom commonly called Tobler's First Law of Geography. If the distances between some features are significantly smaller than the distances between other features, the features are considered to be clustered.

The calculation this tool performs is best done against point data. It can be run against line and polygon data, but the centroids of the features are used. Not only does this type of calculation not represent the data well, but the hypothetical distribution is also flawed when it is compared with the real data.

The other thing to be careful of is the area value the tool uses for the calculations. The default is to use the area of a rectangle defined by the data's outlying values. As an alternative, a nonrectangular measured area can be used.

Calculating the nearest neighbor index is best used for equal study areas. The method works best for comparing data that occurs in the same area over time rather than data that occurs in different areas.

In addition to the nearest neighbor index, the tool calculates a z-score. This number represents a measure of standard deviation that can help you decide whether or not to reject the null hypothesis.

Introduction to spatial statistics

Chapters 8 and 9 deal with spatial statistics. No longer are the commands a simple step to make an output such as a buffer or density. This material gets into a higher level of mathematics and begins to delve into the theoretical realm. With spatial statistics, you deal with the mathematical analysis of existing data to predict the possibility that something will or will not happen. The formulas are complex, and understanding the steps can be difficult. But the results are a less subjective way to confirm what the data represents.

The *Esri Guide to GIS Analysis*, volume 2, is the basis for these chapters, and should be referenced for a complete explanation of the formulas used in the spatial statistics tools. The descriptions and charts it contains provide a good background for the tools and explain how the tools fit into the realm of statistical analysis.

The tutorials in chapters 8 and 9 deal with many of the spatial statistics tools available at all license levels of ArcGIS for Desktop software. Understanding when to use which tool, and why, can be confusing, so a general description of each tool's function is given here.

The first four tools are used to answer a global question: What is the probability that the distribution of these features is occurring due to random chance? Remember that even though the tools do not produce any symbolized data, they are important for establishing a statistical foundation, and a statistical confidence level for your analysis.

8-1
8-2
8-3
8-4

▶ **Average nearest neighbor** (clustering by location). The average nearest neighbor index looks at each feature and the single nearest feature, and then calculates a mathematical index. It creates a hypothetical, randomly distributed set of data and calculates the index again. The degree of clustering is measured by how much the index for the real data differs from the index for the hypothetical data. Use this tool to see if the physical locations are closer together than would be expected with a random distribution.

Getis-Ord General G (clustering by value). The General G statistic looks at the similarity of the values associated with the features within a critical distance of each other. Areas in which the values are similar have a strong clustering, whether they are high values or low values. Use this tool to see if there are areas in which similar values are closer together than would be expected with a random distribution.

Multidistance clustering, or Ripley's K function (clustering by location but using multiple features and multiple distances). Similar to the nearest neighbor method, the K function looks at the distance from a feature to a large number of the nearest features to determine a clustering index. It may also be run for multiple distances to see which distance produces the most significant clustering. This method can detect a more subtle clustering effect than can be detected using the nearest neighbor method. Use this tool to determine if physical locations are clustering due to factors beyond the next nearest feature.

Spatial autocorrelation, or global Moran's I (clustering by both location and value). Spatial autocorrelation determines whether there is an underlying geographic clustering of the data based on both location and attribute value. Use this tool when the physical location data has an attribute associated with it that may be influencing the clustering.

The next two tools can help pinpoint the locations of clustering patterns.

Cluster/outlier analysis, or Anselin local Moran's I. The local Moran's I method identifies areas of clustering by location as well as by values that are similar in magnitude. Use this tool when you want a graphic output on the map, and the suspected clustering is due to both location and an attribute associated with the features.

Getis-Ord hot spot analysis, or Gi* (clustering of high and low values). Gi* indicates areas in which values associated with the features are clustering. A positive index shows clusters of high values (hot spots); a negative index shows clusters of low values (cold spots). Use this tool to create a map display of the locations of hot spot and cold spot clusters based on values.

Within this framework are some general terms and values that are derived from the world of statistical analysis. Terms such as *z-score, null hypothesis, confidence level*, and *statistical significance* are described in *The Esri Guide to Spatial Analysis*, volume 2, and in ArcGIS for Desktop Help, under Geoprocessing.

It is also highly recommended that you view the free Esri Virtual Campus presentation on spatial statistics. Go to the training gateway for more information, http://training.esri.com.

For pattern-analysis tools, the null hypothesis states that there is no pattern. When you perform a feature pattern analysis that yields either a very high or a very low z-score, it indicates that it is very unlikely that the observed pattern is the product of a random distribution.

To reject or accept the null hypothesis, you must make a subjective judgment regarding the degree of risk you are willing to accept for being wrong. This degree of risk is often given in terms of critical values and/or confidence level. At a 95 percent confidence level, the critical z-score values are –1.96 and +1.96 standard deviations. If your z-score is between –1.96 and +1.96, you cannot reject your null hypothesis; the pattern exhibited is a pattern that could very likely be the result of a random distribution. However, if the z-score falls outside that range (for example, –2.5 or +5.4), the pattern exhibited is probably too unusual to be random. A statistically significant z-score makes it possible to reject the null hypothesis and proceed to figuring out what might be causing the clustering or dispersion.

Scenario   The Fort Worth Fire Department has again asked for your help with a spatial analysis project. It wants to know if the emergency medical services (EMS) calls for Battalion 2 have a tendency to cluster. If so, it may consider stationing the emergency intensive care units (EICU) at locations near the hot spots. You will use the Average Nearest Neighbor tool to analyze the data and determine the degree of clustering or dispersion. Ultimately, you will decide whether or not to reject the following null hypothesis:

EMS calls for service in February 2015 are randomly distributed across the study area.

Data   The EMS Calls–Feb15 data is the same response data from the previous tutorials, with a definition query built to extract only the calls with an incident type GRP value between 30 and 39. This group represents the EMS calls only.

Also included is a layer with the Battalion 2 boundary. This new layer has only one polygon. It will be used to determine the area of study, and to make sure that the other battalions are not accidentally included in the study. Remember that the Average Nearest Neighbor tool is very sensitive to the size of the study area.

8-1
8-2
8-3
8-4

Examine the data

You will start with the EMS calls for Battalion 2. By doing quick visual analysis as you did earlier in the book, you can see some groupings of features. You will then analyze the data statistically to assess the extent of clustering. The results of this analysis will add a level of credibility to your reports that purely visual analysis sometimes lacks.

1 In ArcMap, open Tutorial 8-1.mxd.

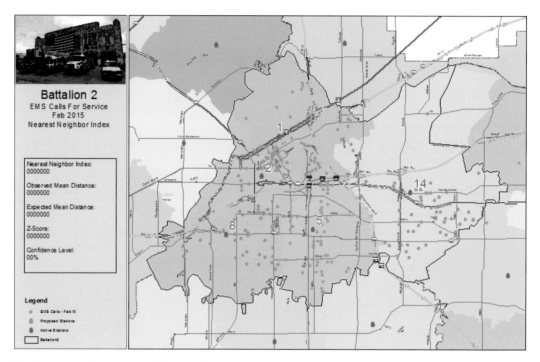

The Average Nearest Neighbor tool is very sensitive to the area of study, so to ensure the best results you will want to look at and record the measured area of the Battalion2 layer.

2 Open the attribute table of the Battalion2 layer and make a note of the square footage in the Shape_Area field.

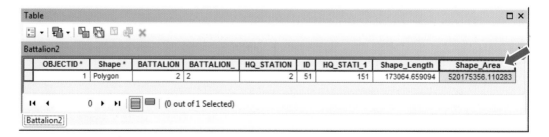

OBJECTID *	Shape *	BATTALION	BATTALION_	HQ_STATION	ID	HQ_STATI_1	Shape_Length	Shape_Area
1	Polygon	2	2	2	51	151	173064.659094	520175356.110283

3 Open the properties of the EMS Calls–Feb 15 layer and examine the definition query that has been created.

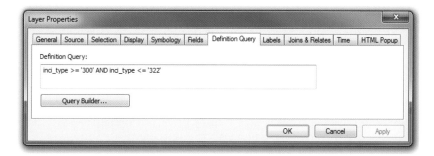

With this information in hand, you are ready to start pattern analysis. This analysis will return a nearest neighbor index and a z-score you will use to decide if the data clusters.

4 Close the Layer Properties dialog box.

Analyze patterns using the Average Nearest Neighbor tool

1 Use the Search window to locate and open the Average Nearest Neighbor tool.

2 Enter the parameters as follows:
   - Input Feature Class: EMS Calls–Feb 15.
   - Area: **520175356** (square feet).
   - Select the Generate Report check box.

3 When your dialog box matches the graphic, click OK.

8-1
8-2
8-3
8-4

4 When the process is complete, click the pop-up box to display the results. If you miss the pop-up box, you can open the Results window from the main menu by clicking Geoprocessing > Results. Make a note of the following values from the results:

- Observed mean distance: _____

- Expected mean distance: _____

- Nearest neighbor ratio: _____

- Z-score: _____

The tool calculates the distance from each feature to its nearest neighbor. Then it finds the average of all these distances. Next, the tool creates a hypothetical random distribution of points using the same number of features and the same study area and repeats the calculations. The option to provide an area ensures that the hypothetical random dataset covers the same area as your study area.

5 In the Results window, double-click the HTML report file to open the graphic display of the results.

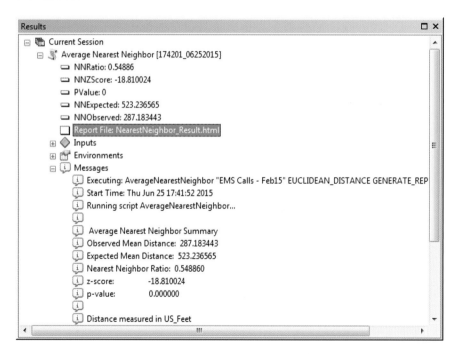

The graphic depicts a bell curve showing the amount of clustering the data displays. It also reports the nearest neighbor ratio (index) and the z-score.

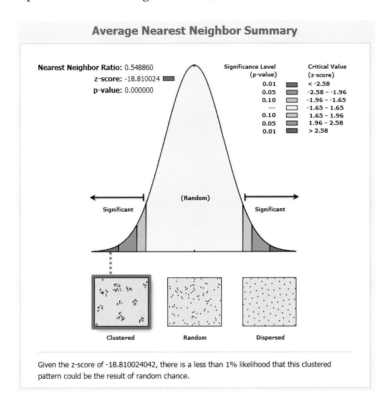

<table>
<tr><th colspan="2" align="center">Average Nearest Neighbor Summary</th></tr>
<tr><td align="right">Observed Mean Distance:</td><td>287.1834 US_Feet</td></tr>
<tr><td align="right">Expected Mean Distance:</td><td>523.2366 US_Feet</td></tr>
<tr><td align="right">Nearest Neighbor Ratio:</td><td>0.548860</td></tr>
<tr><td align="right">z-score:</td><td>-18.810024</td></tr>
<tr><td align="right">p-value:</td><td>0.000000</td></tr>
</table>

Dataset Information

Input Feature Class:	EMS Calls - Feb15
Distance Method:	EUCLIDEAN
Study Area:	520175356.000000
Selection Set:	False

To interpret the results, look at the nearest neighbor ratio, or index, and consider what results the other datasets might produce. A set of features that occurs in a totally random manner would produce an index of 1. A set of features showing more dispersion than the hypothetical data would produce an index greater than 1. The higher the index, the greater the probability that the data is exhibiting statistically significant dispersion. A set of features that shows more clustering than the hypothetical data would produce an index less than 1. The lower the index, the greater the probability that the data is exhibiting statistically significant clustering.

Your dataset returned an index of 0.548860, which means that the features trend toward clustering.

The z-score is –18.810024, with a significance level of 0.01. This score falls into the range that allows you to reject the null hypothesis. If the z-score was closer to 0, it would produce a confidence level that does not support rejecting the null hypothesis. The greater z-score values, either in the positive or negative direction, give a higher confidence level that the distribution is exhibiting clustering or dispersion.

The z-score of –18.810024 gives a 99 percent confidence level that the data distribution is not due to random chance (which is the reverse of the statement in the dialog box that there is less than a 1 percent chance that this clustered pattern could be the result of random chance). With this high confidence level, you can safely reject the null hypothesis.

6 Close the Average Nearest Neighbor graphic display and the Results window.

7 Notice that the Average Nearest Neighbor tool does not create any additional features. It only reports back the values from the requested calculations. To record these results on the map, edit the text box in the title bar and add the results.

Using a box plot

Statisticians will often look at a histogram of the data to see the distribution, as you did in chapter 2. Another standard display of data distribution is the box plot. Box plots can be made using the View > Graphs > Create Graph tool and setting the graph type to Box Plot. Select a layer and the value field, and the graph will be created. The graphics show an example (*left*) of a box plot of the data from this tutorial using the FEE field as the value and a diagram (*right*) of how to read a box plot.

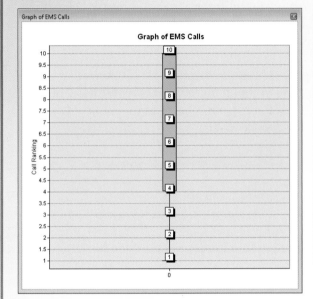

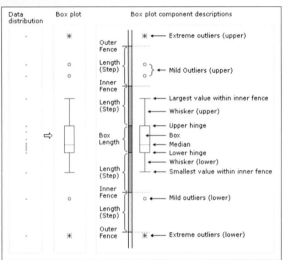

As you work through the other tutorials, try making a box plot to include in your layout.

8 Save your map document as **Tutorial 8-1.mxd** to the \GIST2\MyExercises folder. If you are not continuing to the exercise, exit ArcMap.

8-1
8-2
8-3
8-4

Exercise 8-1

The tutorial showed how to calculate the nearest neighbor index and a z-score. It also explained how to interpret the results.

The fire department now wants to see if false alarms are exhibiting any clustering. If so, the department will target its safety campaign to those areas, explaining how to recognize a real emergency and avoid false alarms.

- Continue using the map document you created in this tutorial, or open Tutorial 8-1.mxd from the \GIST2\Maps folder.
- Turn off the EMS Calls–Feb15 layer.
- Turn on the Incident_Feb15 layer.
- Create a definition query to restrict the data so that the field inci_type is between 700 and 745 inclusively: [inci_type] >= '700' AND [inci_type] <= '745'. Also included are the incident types 7351 and 7352.
- Perform the Average Nearest Neighbor calculations.
- Record the results in the text box in the title bar.
- Save the results as **Exercise 8-1.mxd** to the \GIST2\MyExercises folder.

What to turn in

If you are working in a classroom setting with an instructor, you may be required to submit the maps you created in tutorial 8-1.

Turn in a printed map or screen capture of the following:

Tutorial 8-1.mxd
Exercise 8-1.mxd

Tutorial 8-1 review

The Average Nearest Neighbor tool was used to analyze the fire department data to test for clustering. It returned a set of numbers that gave a confidence level of the statistical analysis. That value, called the z-score, gave a probability of 99 percent that the clustering was statistically significant.

The tool is very sensitive to area. Remember that the nearest neighbor index cannot be compared between areas of different sizes, so it is most valuable for studying results in the same area over time.

Once a set of features is determined to be clustered at a significant level, and the null hypothesis is rejected, you can move on to hot spot or clustering analysis tools.

The Average Nearest Neighbor tool does not use weight fields. The index is affected by only the next nearest feature within the study area.

Study questions

1. Why does an irregular area create problems for the Average Nearest Neighbor tool?

2. Does running the Average Nearest Neighbor tool several times produce the same numbers? Why or why not?

Other real-world examples

A pizza chain might want to determine the clustering of its clients, using the nearest neighbor index. The index will allow the company identify customers who are close to each other, more so than would occur at random, and perhaps categorize the cluster as a delivery hot spot.

An epidemiologist might look for clusters of a certain illness and match the clusters to the locations of food sources or water to determine whether these other geographic elements are associated with an underlying cause.

8-1
8-2
8-3
8-4

Tutorial 8-2

Identifying the clustering of values

With some features, the locations alone are not the factor that determines their clustering. It is instead the values associated with the features. The Getis-Ord General G tool evaluates a value associated with the features being studied and determines whether the high values or the low values cluster.

Learning objectives

- *Test for statistical significance*
- *Evaluate z-scores*
- *Understand random distribution*
- *Understand the clustering of values*

Preparation

- *Read pages 104–8 in* The Esri Guide to GIS Analysis, *volume 2.*

Introduction

The General G statistic measures concentrations of high or low values over the entire study area. It measures how high or low the values associated with the features are within a specific distance of each other, and compares this measurement with how high or low the associated values are in the features outside the specified distance. It is termed *general* because it deals with the entire study area rather than a localized area.

This tool assumes that features have an association because they have similar attribute values, as well as being within the specified distance of each other. The input features for this tool should have an attribute that demonstrates some characteristic or value associated with the location. The result, not the locations, will determine if the values are clustered. Additionally, it will distinguish between clustering of high values versus low values.

Another aspect of this tool is the distance over which the General G statistic is determined to be significant. The "conceptualization of spatial relationships" setting specifies the type of relationships the features have to one another. To set this relationship correctly, you must understand the internal relationships of the data.

If one feature's value has an effect on the values of the nearby features (as with measurements of rain, for example), the inverse distance setting can be used. To amplify this relationship, use the inverse distance squared setting.

For features such as calls for service, it is best to use the fixed distance band setting and run the tool several times to determine how location is affecting the clustering. The value associated with the call does not have an effect on the results because one call for service has no effect on the probability that another call for service will occur in the same area. In other words, a call for service to fight a kitchen fire today will have no relationship to a call for an auto accident on the next block that may happen next week.

Data that has a spatial effect from one feature to another might benefit from a combination of the two settings. The "zone of indifference" setting states that features within a specified distance of each other have an impact on the other features; and after that, the influence quickly drops off. To get the best results, use either the fixed distance band setting, or if you can determine that there is a geographic influence among the features, use the zone of indifference.

The General G tool can be run multiple times for different distances. The highest z-score corresponds to the point of maximum clustering.

Scenario   The fire department wants to know if the calls ranked as a high priority are clustering. Two things that the analysis in tutorial 8-1 did not show were the effect of the FEE ranking on the clustering and the distance at which the clustering occurs. This distance is critical in seeing the compactness of the groupings, and will be used later in the hot spot analysis, in chapter 9.

You will run the analysis at various distances and determine the maximum z-score, confidence level, and distance band for clustering.

Here is the null hypothesis for this analysis:

The Priority Ranking values for the features in Calls For Service–Feb 15 are randomly distributed across the study area.

Based on your analysis, you will decide whether or not to reject the null hypothesis.

Data   The Calls For Service–Feb15 data is the same response data from the previous tutorials. The field FEE contains ranks from 1 to 10, reflecting each call's priority.

Additional layers are included for background interest.

8-1

8-2

8-3

8-4

Investigate threshold distances

1 In ArcMap, open Tutorial 8-2.mxd.

Once again, here is a map of the Battalion 2 calls for service. Even though you might see some groupings, you cannot see what effect the call priority has. Visual analysis cannot provide that result, so you must instead rely on spatial statistics for the answer.

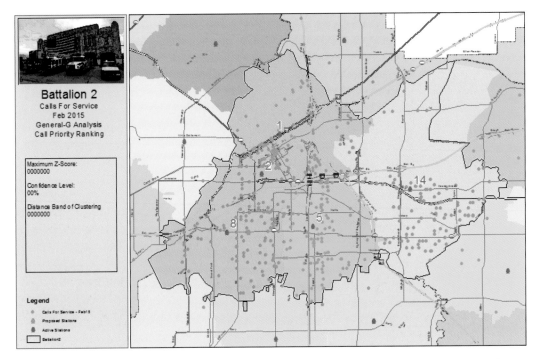

You will run the General G tool several times to find the value at which the z-score peaks, but what values should you try? If you do not get within the right range, you may not discover the clustering. To find the distance range to try, you will use a combination of visual analysis and a statistics utility.

2 Zoom in to an area that exhibits some visual grouping. A quick scan shows about seven or eight calls located in this area.

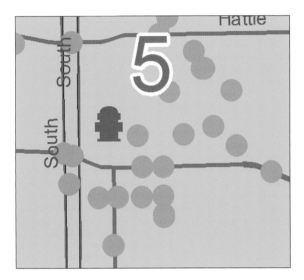

You will use the Calculate Distance Band from Neighbor Count and the observed number of neighbors from your visual inspection to get an idea of what distance to enter in the General G tool. Then you can start recording z-scores.

3 Locate and open the Calculate Distance Band from Neighbor Count tool.

4 Use Calls For Service–Feb 15 as the input features, and set the number of neighbors to **7**. Click OK.

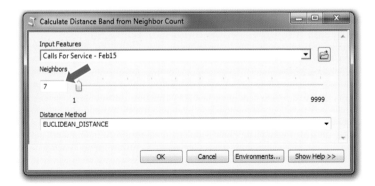

The tool finds the minimum, average, and maximum distances at which each feature can find at least seven neighbors.

8-1
8-2
8-3
8-4

5 Open the Results window when the process is complete; and note the minimum, average, and maximum distance. You will base your search distances on the average distance. Close the Results window.

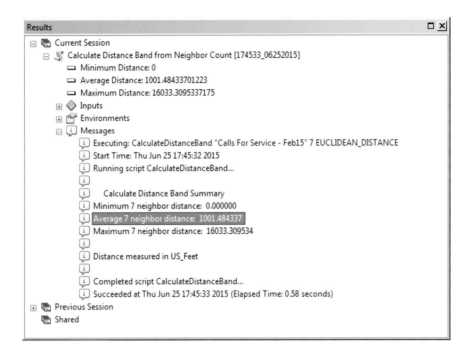

The average distance to find seven neighbors is around 1,100 feet. To determine which values to use with the General G tool, use numbers that range both above and below the results of your initial investigation. The distances to try in the General G tool should be from 1,000 feet to 1,400 feet, at 100-foot intervals. Because the map units are feet, these numbers can be entered without adding units. But if the map units are set to something else, you can enter the abbreviation for feet in the input box. Make a note of the z-score at each of the distances, and watch for the peak value in this range. The distance that produces the largest z-score will be the distance with the most significant clustering.

Run the High/Low Clustering (Getis-Ord General G) tool

1 Locate and open the High/Low Clustering tool.

2 Enter the parameters as follows:
- Input Feature Class: Calls For Service–Feb 15.
- Input Field: FEE.
- Select the Generate Report check box.
- Conceptualization of Spatial Relationships: FIXED_DISTANCE_BAND.
- Distance Band: **1000**.

3 When your dialog box matches the graphic, click OK.

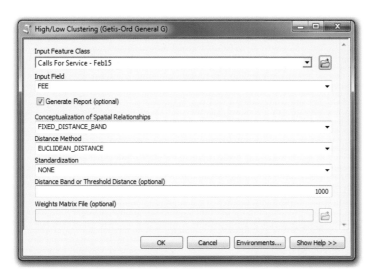

4 Double-click the HTML report file in the Results window to view the graphic report.

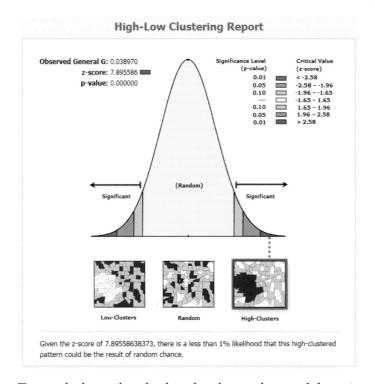

The result shows that the data does have a degree of clustering at this distance. But is this the distance at which the greatest clustering is found? Also note that many features do not have neighbors at this distance, which makes them invalid in the calculation. A better distance would have far fewer features with no neighbors so that the maximum number of features are included in the calculation.

5 Record the General G index and the z-score for 1,000 feet. Then close the Results window and the clustering report.

YOUR TURN

Run the High/Low Clustering tool five more times, increasing the distance by 200 feet each time. Record the G indexes and z-scores for distances of 1,000; 1,100; 1,200; 1,300; and 1,400 feet. You might want to list your results in a table, as shown.

Distance	G index	Z-score
1,000		
1,100		
1,200		
1,300		
1,400		

Graphing the results will show where the z-score peaks, which is the point of a significant clustering of high values. Remember that the z-scores show the distance at which values are clustering at a rate higher than expected by chance. Notice also that as the distance increases, the number of features with no neighbors decreases.

6 Update the text box in the title bar with the maximum z-score, confidence level, and distance band of clustering.

7 Save your map document as **Tutorial 8-2.mxd** to the \GIST2\MyExercises folder. If you are not continuing to the exercise, exit ArcMap.

Exercise 8-2

The tutorial showed how to calculate a z-score to assess whether the values of features are clustered. The fire department wants you to do the same clustering analysis for another set of data.

- Open Tutorial 8-2E.mxd from the \GIST2\Maps folder.
- Perform the high/low clustering calculations for the Calls For Service–Jan 15 layer. (**Hint:** Use distances from **700** to **1,200** feet.)
- Record the z-scores for each distance and determine which distance provides the highest z-score.
- Record the results in the text box in the title bar.
- Change the titles, colors, and legend to make a visually pleasing map.
- Save the results as **Exercise 8-2.mxd** to the \GIST2\MyExercises folder.

What to turn in

If you are working in a classroom setting with an instructor, you may be required to submit the items you created in tutorial 8-2.

Turn in a printed map or screen capture (together with your z-score tables) for the following:

Tutorial 8-2.mxd
Exercise 8-2.mxd

8-1
8-2
8-3
8-4

Tutorial 8-2 review

The values for the features were used to demonstrate clustering. By running the tool several times with a range of distances, the highest z-score is found. This peak z-score corresponds to the distance at which the clustering of values is the strongest.

Study questions

1. What other types of values might be used for this analysis?

2. How does this analysis differ from the average nearest neighbor analysis?

Other real-world examples

A retail chain might determine the clustering of customers who have spent a large amount of money in the stores to set up a home-based shopping experience.

A map of crop yield might be used to determine clustering of higher-than-expected harvesting numbers. This analysis may in turn be used to study factors contributing to high growth.

Oceanographers may look for clustering of measured salinity levels in the ocean and match the results to currents to see if there is any correlation.

Tutorial 8-3

Checking for multidistance clustering

Multidistance spatial cluster analysis, also known as Ripley's K function, examines the counts of neighboring features at several distances. If the count is higher than what would occur in a random distribution, the features are considered clustered. A graph is automatically created for the tested distances.

Learning objectives

- *Test for statistical significance*
- *Evaluate z-scores*
- *Understand random distribution*
- *Evaluate clustering of values at multiple distances*

Preparation

- *Read pages 97–103 in* The Esri Guide to GIS Analysis, *volume 2.*

Introduction

Using the Ripley's K function for pattern analysis measures the distance between features to determine clustering. But unlike the average nearest neighbor index, it includes all the neighboring features in the calculation, not just the nearest one.

As with other methods, the Ripley's K function generates a hypothetical random distribution using the same number of features and the same area. The calculation is done on both the real data and the hypothetical data, and the difference between the observed index value and the index value generated by the hypothetical random data indicates the degree of clustering. When plotted on a graph, the area with the greatest distance between the two values is the point of most significant clustering.

The index is calculated by measuring the distance from each feature to all the other features in the dataset; then a mathematical operation generates a K value. The process is repeated for various user-specified distances and plotted on a graph. Then the tool creates a hypothetical set of data by randomly tossing the same number of features into the same study area and calculating the index again. You enter a start distance and an incremental distance for the tool to run through a range of distances, eliminating the need to run the tool multiple times and manually plot the results.

Using the optional permutations, the calculation is performed against the random dataset multiple times, creating a confidence envelope. The highest and lowest values reached in the random datasets that are created in the multiple permutations are represented on the graph. The peak at which the observed K index exceeds the confidence envelope the most is the point at which the greatest clustering occurs.

Because the distances between all the features in the dataset are used in the Ripley's K function calculation, the geographic shape of the area has a great impact on the results. The hypothetical random dataset is created in the same area, and must be subject to the same shape and distance as the real data. It is recommended that a single polygon representing the study area be used in the calculations.

Another feature of multidistance clustering is the option to include a weight. As with the other tools, the weight value will give some features more importance in the calculations.

Scenario

The pattern analysis tools you have used proved the response call dataset to have significant clustering. The Multi-Distance Spatial Cluster Analysis tool will take into account the specific study area, and include the relationships of each feature to all the other features. Using this tool, along with the weighting factor, will produce a better result. You will run the analysis for a set of distances and determine the maximum z-score, confidence level, and distance band for clustering.

Here is the null hypothesis for this analysis:

The features in the Calls For Service–Feb 15 dataset, when weighted with priority rankings, are randomly distributed across the study area.

Based on your analysis, you will decide whether or not to reject the null hypothesis.

Data

The Calls For Service–Feb15 data is the same response data from tutorials 8-1 and 8-2. A field containing FEE values contains ranks from 1 to 10, reflecting each call's priority.

Additional layers are included for background interest.

Run the Multi-Distance Spatial Cluster Analysis (Ripley's K function) tool

1   In ArcMap, open Tutorial 8-3.mxd.

Once again, here is a map of the Battalion 2 calls for service. You will use Ripley's K function to analyze the patterns of this dataset, in relation to the unique shape of Battalion 2. The features that fall outside Battalion 2 will have an effect on the data, both by the distance they are from the other features and their weight values.

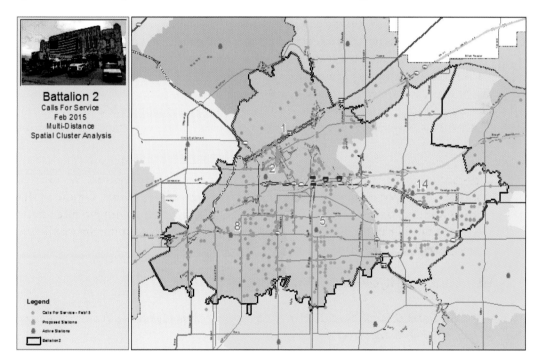

2   Locate and open the Multi-Distance Spatial Cluster Analysis tool.

The tool has many parameters and options to set, and each one can have a large impact on the results. The ones used here are based on the distances used for the high/low clustering calculations and will encompass the value that showed the best clustering.

8-1
8-2
8-3
8-4

3 Set Input Feature Class to Calls For Service–Feb 15, and set the location for the output table to \GIST2\MyExercises\MyData**KFunctionFeb15**.

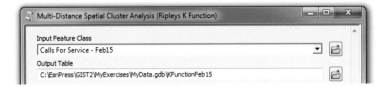

The output table will store all the various outcomes of the calculations. After the tool completes the calculations, you can use this data to make a graph to include on the map.

4 Fill out the next portion of the dialog box as follows:
- Number of Distance Bands: 10.
- Select the Display Results Graphically check box.
- Weight Field: FEE.
- Beginning Distance: **200**.
- Distance Increment: **100**.

The number of distance bands determines how many measurement rings to make. The distance and increment values give a start value and a value for how far apart each ring will be, respectively. With the values you entered, the first ring will be at 200 feet, and then nine more rings 100 feet apart will be used. As with other analyses, the calculation will be weighted based on FEE values, giving call priority ranking an influence on the results. The optional confidence envelope is left at zero for now, but you will use it later in the tutorial.

5 Set Boundary Correction Method to NONE. Finally, set Study Area Method to "User-provided study area feature class" and Study Area Feature Class to Battalion 2.

You can use the boundary correction method to help simulate features falling outside the study area. This method can give the features that are nearest the edge simulated neighbors to prevent a black hole effect. This dataset has features that fall outside the study area, so no boundary correction is necessary. If the dataset is cut at the boundary from a larger dataset, a boundary correction is appropriate.

As mentioned before, this tool is very sensitive to the study area provided, especially if multiple permutations are used. The polygon that represents Battalion 2 was selected and exported to a separate feature class to use with this calculation. Setting the study area method to "User provided" allows this feature class to be used for all the calculations.

6 **Click OK. When the process finishes, open the Results window.**

The results are displayed in the Results window, and all the values are stored in a table called KFunctionFeb15, which is added to your table of contents.

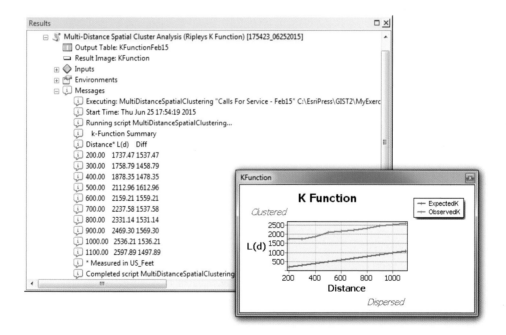

7 Close the Results window.

Create a graph of the Difference field

The distance of the most significant clustering will be where the observed values exceed the expected values by the greatest margin. You can show this difference in a graph.

1 At the top of the table of contents, click List by Source. Right-click KFunctionFeb15 and click Open to display the table. Click Table Options and then click Create Graph.

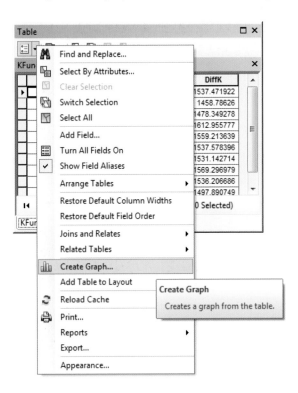

2   In the Create Graph wizard, set Graph type to Vertical Line, set Y field to DiffK, and set X label field to ExpectedK. When your dialog box matches the graphic, click Next. Change the title to DiffK Values and click Finish to create the graph.

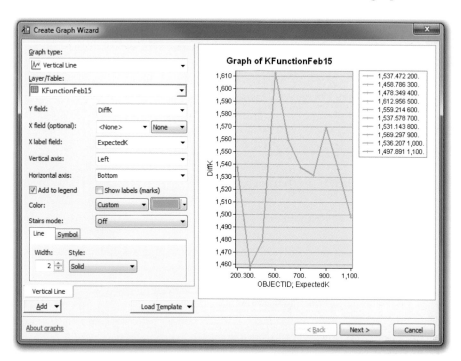

3   Right-click the graph and click Add to Layout. Close the graph and the attribute table.

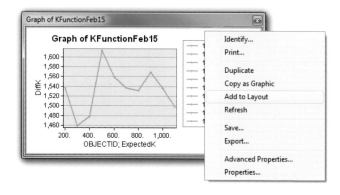

4 Resize the graph in the layout and move it to the title bar.

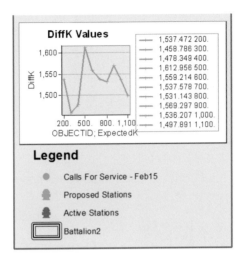

The graph shows a peak in the values at 900 feet, but there is another peak at 500. Either of these peaks might be the point of the most pronounced clustering.

This set of calculations does not include the confidence envelope that the permutations option provides. The confidence envelope is developed by running hypothetical random distributions many more times, and plotting the highest and lowest values for each distance. You can then compare the values of the Difference field to a larger sampling of possible random data. The value that exceeds the confidence envelope by the greatest margin is the distance of most significant clustering.

Rerun the Multi-Distance Spatial Cluster Analysis tool using a confidence envelope

1 Open the Multi-Distance Spatial Cluster Analysis tool.

2 Set the following parameters:
- Input Feature Class: Calls For Service–Feb15
- Output Table: \MyExercises\MyData.gdb\KFunctionFeb15ConfEnv
- Number of Distance Bands: 10
- Compute Confidence Envelope: 99 Permutations

Note: this many permutations may take one to two minutes to complete. If you have time, try also using 999 permutations and see what difference it might make in the results.

- Select the Display Results Graphically check box.
- Weight Field: FEE.
- Beginning Distance: **200**.
- Distance Increment: **100**.
- Boundary Correction Method: NONE.
- Study Area Method: USER_PROVIDED_STUDY_AREA_FEATURE_CLASS.
- Study Area Feature Class: Battalion 2.

3 When your dialog box matches the graphic, click OK.

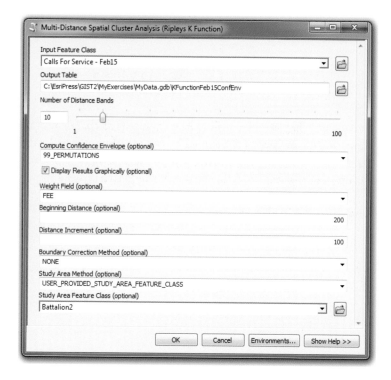

The tool calculates the *K* index for the real data and then calculates the index for the random hypothetical data, each 99 times. Comparing this expanded set of data with the real value will make it easier to see which clusters are unlikely to occur randomly.

4   Note the results of the K function and then close the Results window. A table that contains the results is added to the table of contents (on the List by Source pane).

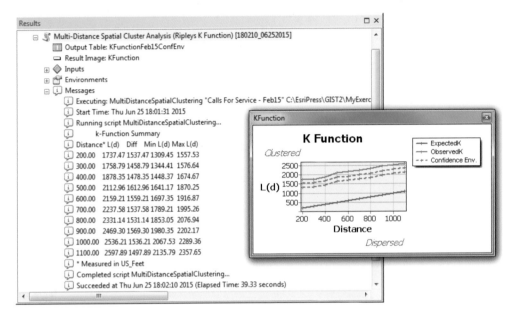

Examine the results

To find the distance of maximum clustering, it is best to calculate the difference between the observed *K* index and the upper limit of the confidence envelope.

1 Open the KFunctionFeb15ConfEnv table. Create a new floating point field named
Difference in the table and calculate the following value: [ObservedK] – [HiConfEnv].
Click OK.

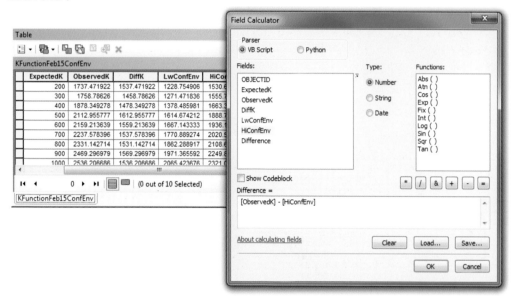

2 Make a graph as you did earlier in this chapter, using Difference as the Y field and
ExpectedK as the X label field. (Your results may differ depending on the number of
permutations used.)

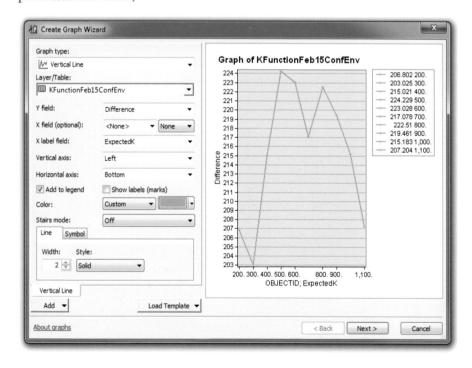

3 Add the graph to the layout, and then close the graph and attribute table.

4 Resize the graph and add it to the title bar.

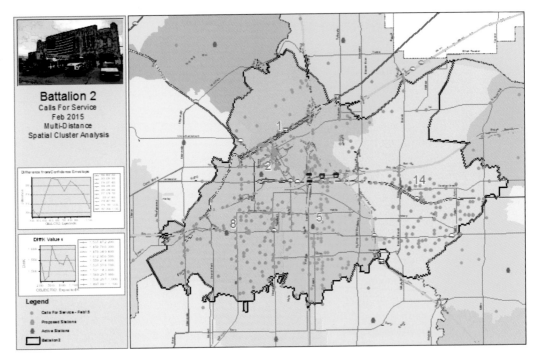

The resulting graph makes it clear that the 500 value is the peak of the graph. The difference between the two peaks is the use of the confidence envelope and the 99 permutations to check a larger sample of the hypothetical random data.

5 Save your map document as **Tutorial 8-3.mxd** to the \GIST2\MyExercises folder. If you are not continuing to the exercise, exit ArcMap.

Exercise 8-3

The tutorial showed the use of the Multi-Distance Spatial Cluster Analysis tool for calculating a K function index and a confidence envelope to assess the significance of clustering.

The fire department wants you to do the same analysis for another set of data.
- Open Tutorial 8-3E.mxd from the \GIST2\Maps folder.
- Perform the multidistance cluster analysis calculations using a confidence envelope for the Calls For Service–Jan15 layer. Use the same distance parameters as in the tutorial.
- Create a graph from the data and add it to the map.
- Change the titles, colors, and legend to make a visually pleasing map.
- Save the results as **Exercise 8-3.mxd** to the \GIST2\MyExercises folder.

What to turn in

If you are working in a classroom setting with an instructor, you may be required to submit the items you created in tutorial 8-3.

Turn in a printed map or screen capture of the following:
> **Tutorial 8-3.mxd**
> **Exercise 8-3.mxd**

Tutorial 8-3 review

The spatial statistics tools from tutorials 8-1 and 8-2 use a standard formula for computing the z-score, and they can be run many times to test the data. The Ripley's K function uses permutations to run many scenarios, basing the results on many more tests of random hypothetical data.

Because so many random samples are used, the area in which they are generated is very critical to making the hypothetical data as similar to the real data as possible. If the random data is distributed in a much larger space, it will not simulate reality well, and will give misleading results.

Finding the distances to use for the tool can be rather subjective. If you have detailed knowledge of the data, and know some of the distances that may affect the data, these distances might serve as the basis for the analysis. For instance, if you are looking for clusters of commute distances and know that the average commute is 15 miles, this mileage may guide you in selecting distances.

The data can cluster at different distances. The confidence envelope provides a basis for comparing the levels of clustering and determining what level is the most significant.

Study questions

1. How is the confidence envelope calculated?

2. What does the confidence envelope tell you?

3. Do you expect to see a difference between 9 and 999 permutations? Why?

Other real-world examples

Ripley's K function can be used to examine the clustering of a certain plant species. A regional search may be done for the significance of clustering rather than using only the next nearest neighbor.

A catalog sales company may look for clustering of customers but on a more regional level. This clustering will take into account other customers across a broader distance and reveal more subtle patterns.

Tutorial 8-4

Measuring spatial autocorrelation

Measuring spatial autocorrelation involves not only using feature locations or attribute values alone, but using both simultaneously. Given a set of features, a Moran's I index is calculated to determine if the data is clustered, random, or dispersed. Moran's I requires a weight from either an attribute or a count of features in a defined area.

Learning objectives

- *Test for statistical significance*
- *Evaluate z-scores*
- *Understand random distribution*
- *Evaluate clustering of values and distances*

Preparation

- *Read pages 80–87 and 118–26 in* The Esri Guide to GIS Analysis, *volume 2.*

Introduction

The Moran's I function is designed to find clustering using the attribute values as well as the locations of features. The function is typically used with polygons that contain a summary statistic, such as census data or density data. It is important to note that the Moran's I function does not identify clusters on the map, but rather identifies whether the pattern of values across the study area tends to be clustered, random, or dispersed.

The point data of calls for service will be aggregated into polygons for this analysis. The polygons will provide clusters of density. One way to create summary polygon data is to overlay a grid of a set size and count the number of locations that fall within each polygon. ArcMap contains a command to create the grid; then you can perform a spatial join to count the features in each cell.

Determining cell size is critical to the analysis. Because the tool looks for clustering of densities, it is important to choose a grid size that will include polygons that have at least one point in them. The grid must also provide a larger range of values across the dataset. If you have some idea of the dataset's distance characteristics, you can use this knowledge to determine a grid size distance. For the calls for service data, for instance, using the size of a standard city block, or 500 feet, is acceptable.

Once the grid is created and the spatial join is performed, the Moran's I tool is run using the output grid as the features and the count of call locations as the attribute value.

The Moran's I tool compares the values for neighboring features. A comparison is made of the differences in values between each pair of neighbors and all the other features in the study area. If the average difference between neighboring features is less than between all the features, the values are considered clustered.

As with the other spatial statistics tools, the most significant clustering occurs at the point where the z-score peaks. The command can be run several times to find the peak z-score.

Scenario
The fire department is looking at density of calls per block and wants to see the distance at which these densities cluster.

The first step is to overlay a grid onto the point data and count the number of calls in each cell. Then the Moran's I tool can be run with a range of distances to determine at what point the features cluster.

Here is the null hypothesis for this analysis:

Calls for service are randomly distributed among city blocks.

Data
The Calls For Service–Feb 15 data is the same response data from the previous tutorials in this chapter. Additional layers are included for background interest.

Aggregate the data

1 In ArcMap, open Tutorial 8-4.mxd.

Once again, here is a map of the Battalion 2 calls for service. A grid will be overlaid and used to count the features, creating a density grid. Then you can use the Spatial Autocorrelation Moran's I tool to look for clustering of the density values.

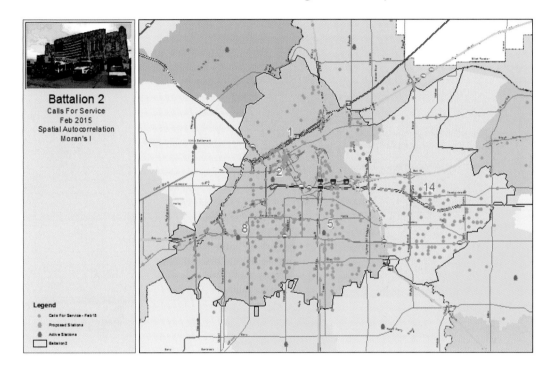

2　Turn on the layer 500 ft Grid.

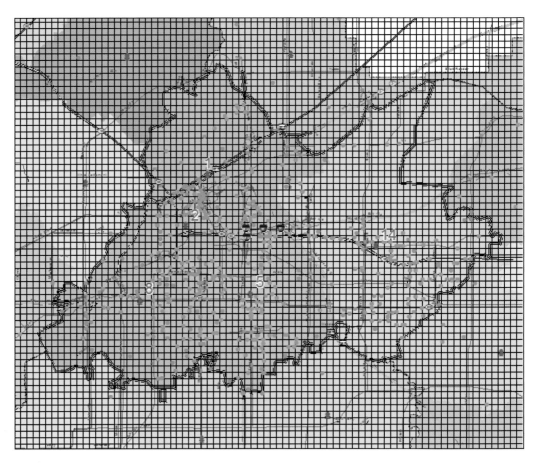

The layer is a 500-by-500-foot grid. Creating a spatial join between the grid layer and the Calls For Service layer will output a layer with the same grid and a count of how many call locations occur inside each cell.

3　Right-click the 500 ft Grid layer and click Joins and Relates > Join. Complete the Join Data dialog box as follows:

- Under "What do you want to join to this layer?" select "Join data from another layer based on spatial location."
- For item 1, set the join layer to Calls For Service–Feb15.
- For item 3, set the output file to \MyExercises\MyData.mdb\ Calls_for_Service_Feb15_Grid.

4 When your dialog box matches the graphic, click OK.

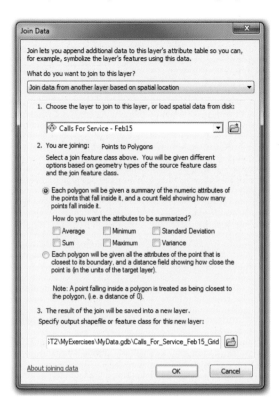

The polygon file that was created now shows the density of the calls for service. Before running the Spatial Autocorrelation tool, you will remove the cells that have no value. You can remove the cells using a definition query.

5 Turn off the 500 ft Grid layer.

6 Open the properties of the Calls_For_Service_Feb15_Grid layer and create a definition query to allow only Count values that exceed zero. Click OK and then OK again.

8-1
8-2
8-3
8-4

The result displays only the cells that contain at least one call for service. Although you may look at these cells and think you see clustering, you cannot see the values associated with the cells that will be used in the spatial autocorrelation calculation.

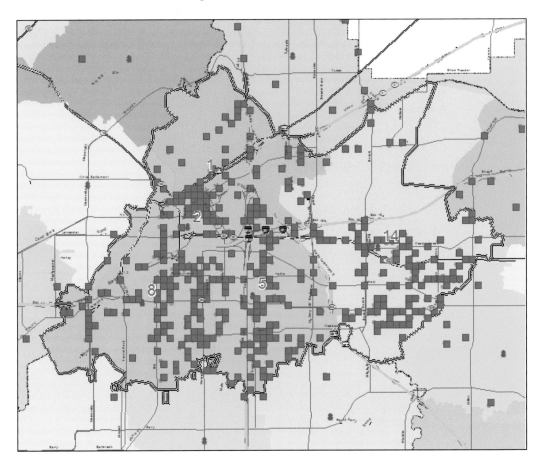

Run the Spatial Autocorrelation tool

The Spatial Autocorrelation tool uses familiar settings from the other spatial statistics tools. You will enter input data to analyze. Then you will select a value field and a conceptualization of spatial relationships method. You will run the tool for a range of values to determine the peak z-score. With the grid set at 500 feet, the search for the peak z-score can begin with grids of at least seven neighbors—a distance of around 3,400 feet. Use a test-distance increase of 200 feet for each trial until the z-score peaks.

1   Locate and open the Spatial Autocorrelation tool.

2   Enter the following parameters in the dialog box:

- Input Feature Class: Calls_For_Service_Feb15_Grid.
- Input Field: Count.
- Select the Generate Report check box.
- Conceptualization of Spatial Relationships: ZONE_OF_INDIFFERENCE.
- Distance Band: **3400** (feet).

3   When your dialog box matches the graphic, click OK.

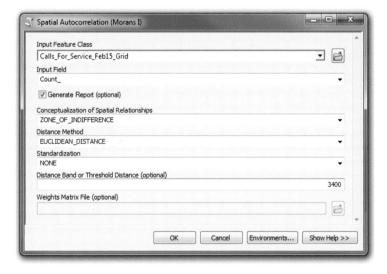

4 Record the resulting z-score from the Results window.

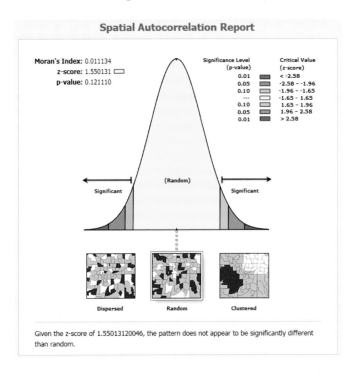

Spatial Autocorrelation Report

Moran's Index: 0.011134
z-score: 1.550131
p-value: 0.121110

Significance Level (p-value)	Critical Value (z-score)
0.01	< -2.58
0.05	-2.58 – -1.96
0.10	-1.96 – -1.65
---	-1.65 – 1.65
0.10	1.65 – 1.96
0.05	1.96 – 2.58
0.01	> 2.58

Significant   (Random)   Significant

Dispersed   Random   Clustered

Given the z-score of 1.55013120046, the pattern does not appear to be significantly different than random.

YOUR TURN

Repeat the process for the distances shown in the table. Keep all the other parameters the same, changing only the distance band. You might want to list your results in a table, as shown.

Examine the z-scores and find the highest one. Check the confidence level to determine if the block-level data is clustered. Based on the results, can you reject the null hypothesis?

Distance	Z-score
3,400...............	
3,600...............	
3,800...............	
4,000...............	
4,200...............	
4,400...............	
4,600...............	

5 Save your map document as **Tutorial 8-4.mxd** to the \GIST2\MyExercises folder. If you are not continuing to the exercise, exit ArcMap.

Exercise 8-4

The tutorial showed the use of the Spatial Autocorrelation tool to determine Moran's I values and z-scores to assess the level of clustering of the polygon data.

Repeat the analysis using the patron location data from the Oleander Library. Because the patrons are point data, you will aggregate this data to polygon data before running the autocorrelation tools. A 300-foot grid has been provided for this exercise, but you can create your own grid using the Create Fishnet tool.

- Open Tutorial 8-4E.mxd.
- Perform a spatial join of the 300-foot grid using Patron Locations.
- Create a definition query as you did in the tutorial.
- Determine the distance to use in the analysis. (**Hint:** nine or 10 neighbors on a 300-foot grid represents 2,800 feet.)
- Run the Spatial Autocorrelation tool for the new grid layer using your range of test distances.
- Record the z-scores and distances with their confidence levels and determine whether you can reject the null hypothesis. Show the most significant results in the text box in the title bar.
- Change the titles, colors, and legend to make a visually pleasing map.
- Save the results as **Exercise 8-4.mxd** to the \GIST2\MyExercises folder.

What to turn in

If you are working in a classroom setting with an instructor, you may be required to submit the items you created in tutorial 8-4.

Turn in a printed map or screen capture, with z-scores, for the following:

Tutorial 8-4.mxd
Exercise 8-4.mxd

8-1
8-2
8-3
8-4

Tutorial 8-4 review

Moran's I measures the clustering of value densities. The locations of features and their values are used in the calculations. As with the other pattern analysis tools, the z-score is tracked as a measure of confidence in the clustering. The fire department calls for service data used in the tutorial was tested against a 500-foot grid to see if the data clusters at the block level. The z-scores show that any apparent clustering is probably due to chance.

For point data for which there is no grid available, it is possible to aggregate the data to another point feature class. You can use the Integrate command to group several of the points into a single location, keeping a count of how many features are integrated. The result is a point feature including summarized data that can be analyzed using the Spatial Autocorrelation tool.

Study questions

1. What, if anything, might be changed to find significant clustering in the data?

2. What two factors influence the results of Moran's I?

3. Why do the other pattern analysis tools find clustering in this data, whereas the Moran's I does not?

Another real-world example

The Moran's I may be used to examine gas-well production levels. Both the density and output of the wells are important in the calculation. Wells that cluster with high production might suggest that more wells in that same area will also be productive.

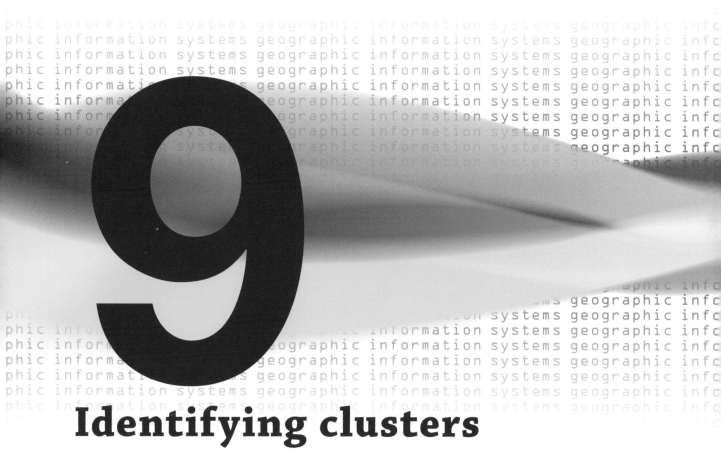

9

Identifying clusters

Once you establish the statistical significance of patterns in a dataset, you can then pinpoint the locations of clustering patterns. The two methods discussed in this chapter apply statistical values identified using the pattern analysis tools to produce a visual output that can then be used to graphically display the results. You can use the results to analyze the data's historical significance or help predict future trends.

Tutorial 9-1

Performing cluster and outlier analysis

Using pattern analysis tools to check z-scores provides the basis for determining the probability that spatial clustering is not due to random chance. You can then use cluster and outlier analysis to display the results graphically. The Cluster and Outlier Analysis tool (Anselin local Moran's I) stores the results of the pattern analysis in the layer's attribute table. You can then use these results in symbol classification.

Learning objectives

- *Analyze spatial clustering*
- *Evaluate z-scores*
- *Identify hot spots and cold spots*
- *Classify symbols*

Preparation

- *Read pages 147–74 in* The Esri Guide to GIS Analysis, *volume 2.*

Introduction

Pattern analysis allows you to determine if clustering is occurring at a high enough probability to reject the null hypothesis. Once it is determined that data is clustering, you can examine the causes of the clustering by displaying the results graphically.

The Cluster and Outlier Analysis tool takes a set of features using a weight or attribute value, performs pattern analysis, and displays the results in such a way as to highlight the clustering. Areas in which high values cluster are called *hot spots*, and areas in which low values cluster are called *cold spots*. When these patterns are displayed graphically, they can be overlaid on other datasets to help identify contributing factors.

The Cluster and Outlier Analysis tool calculates several new values that are used to analyze the data. These values are added to new fields in the output feature class. The LMiIndex indicates whether the feature has other features of similar values near it. Positive numbers represent features that are part of a cluster, and negative values represent features that are spatial outliers. The LMiZScore shows the statistical significance of the clustering, as described earlier, and the LMiPValue is the probability of error in rejecting the null hypothesis. The COType field distinguishes between a statistically significant (0.05 level) cluster of high values (HH), cluster of low values (LL), an outlier high value surrounded primarily by low values (HL), and an outlier low value surrounded primarily by high values (LH).

Scenario   After running several spatial pattern analyses on the call for service data, the fire chief needs a map that shows the results for the next city council meeting. The chief also wants to show hot spots and cold spots based on feature locations and the ranking of the calls. Then the chief can overlay the data on the census data to see if there is some type of relationship between the two datasets.

Data   The first dataset is Calls for Service data that the fire department compiled for February 2015. In addition, there is an attribute field that contains a number from 1 to 10 to designate the priority of the call.

The second set of data is the census block-group-level data. It contains a field for total population, which can be used in a quantity classification. The area you need is already cut out for use, but the analysis can be repeated for any area using the exercise data.

The third set of data is the street network data to give the map some reference. It comes from the Data and Maps for ArcGIS data.

Run the Cluster and Outlier Analysis tool

1   In ArcMap, open Tutorial 9-1.mxd.

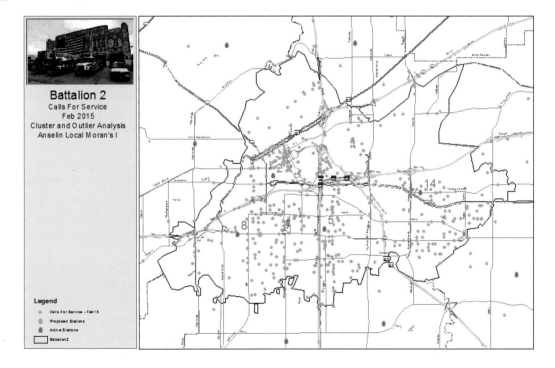

Here is the familiar Battalion 2 data. First, you will run the Local Moran's I Analysis tool to show where the calls for service might be clustered. Then you will display the census data with it to see if there is a relationship between the two.

2 Use the Search window to locate and open the Cluster/Outlier Analysis with Rendering tool.

3 Enter the parameters as follows:
- Input Feature Class: Calls for Service–Feb15
- Input Field: FEE
- Output Layer File: \GIST2\ MyExercises\MoransICluster.lyr
- Output Feature Class: \GIST2\MyExercises\MyData.gdb\MoransICluster

4 When your dialog box matches the graphic, click OK.

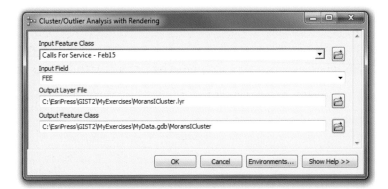

When the tool finishes, it creates a feature class with the results of the Moran's I analysis. The results include values for the LMiIndex, LMiZScore, and LMiPValue, as discussed in the introduction of this chapter.

5 Open the attribute table of the Moran's I layer and note the new values that were added. When you are finished, close the table.

	Table				□ ×

MoransICluster

SOURCE_ID	FEE	LMiIndex IDW 10037	LMiZScore IDW 10037	LMiPValue IDW 10037	COType IDW 10037
3	2	-3.652889	-1.828305	0.067504	
4	6	0.000009	0.069935	0.944245	
5	6	-0.002797	-0.530387	0.595844	
6	8	13.529167	2.199185	0.027865	HH
7	6	-0.20536	-0.116626	0.907157	
8	2	0.012231	3.098021	0.001948	LL
9	8	0.724701	0.206898	0.83609	
11	8	13.448476	1.699701	0.089187	
12	8	13.448476	1.699701	0.089187	

I◄ ◄   1   ► ►I   (0 out of 846 Selected)

MoransICluster

In the newly symbolized feature class, the red dots represent hot spots, or areas where features have high positive z-scores because they are surrounded by features with similar values, either high or low. The blue dots represent the cold spots, or areas where features have low negative z-scores because they are surrounded by features with dissimilar values.

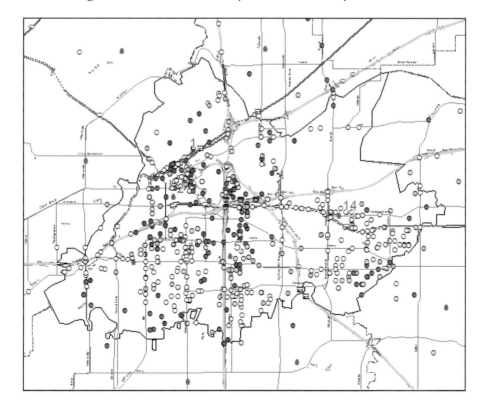

Because high z-scores can represent both high and low values of the input field, it will help to know where high values are surrounded by other high values, or low values are surrounded by other low values. An additional field is added to the results table to show where these patterns occur.

6 Create a copy of the layer MoransICluster in the table of contents and rename it **ClusterTypes**. Set the symbology to Unique Values and the Value field to COType IDW 10037.

7 Add the values HH, HL, LH, and LL. Clear all other values.

8 Change the symbols to Circle 1 with a size of 50. Change the color of HH to Fire Red, LL to Leaf Green, and both HL and LH to Solar Yellow.

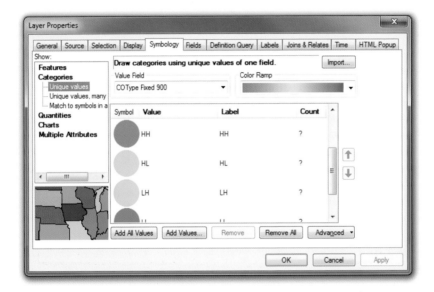

9 Set the transparency of the layer to 40%. Close Layer Properties.

10 Move the ClusterTypes layer below the MoransICluster layer.

The red symbols show where high values are clustering with other high values. The green symbols show where low values are clustering with other low values. The yellow symbols show where high and low values are clustering. Areas in which there is not significant clustering of values do not get a COType value and therefore do not appear on this map.

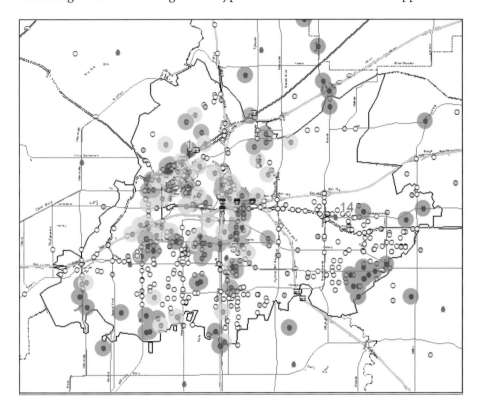

The hot spots and cold spots are shown, but what is the cause of this phenomenon? One reason might be that the areas with the highest density of population cause the most calls for service.

11 Turn on the Census 2010 layer.

Look for areas with hot spots or cold spots and see if you can determine a relationship with the census data. Some hot spots occur in areas with a high population, and some occur in areas with a medium population. The cold spots seem to occur in all the population ranges. After visually assessing the data, it seems that there is no correlation between these two datasets. Further investigation into other data layers might provide better insight into any underlying causes of the clustering.

Rerun the Cluster and Outlier Analysis tool using a fixed-distance band

The Moran's I tool with rendering uses inverse distance for the conceptualization of spatial relationships. Inverse distance is not always the best choice because you are not able to control the search distance it uses to find a significant number of neighbors. In the tutorials throughout chapter 8, you plotted the z-score for various distances to determine which z-score produced the best results in regard to nearby values. You will run the Moran's I tool again without the rendering and set a fixed-distance band according to what you discovered before in using the pattern analysis tools.

1 Turn off the Census 2010, ClusterTypes, and MoransICluster layers.

2 Locate and open the Cluster and Outlier Analysis (Anselin Local Moran's I) tool (use the tool without rendering).

3 Enter the parameters as follows:
- Input Feature Class: Calls for Service–Feb15
- Input Field: FEE
- Output Feature Class: \GIST2\MyExercises\MyData.gdb\MoransICluster900
- Conceptualization of Spatial Relationships: FIXED_DISTANCE_BAND
- Distance Band: **900** feet

4 When your dialog box matches the graphic, click OK.

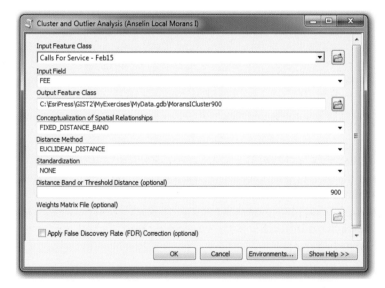

The resulting feature class is added to the table of contents, and several new fields are added to the attribute table. The field LMiIndex has the resulting Moran's I index value. You can symbolize this layer the same as the first layer you created to show hot spots and cold spots. Remember that positive values show clustering, and negative values show dispersion.

5 Open the properties of the MoransICluster900 layer and duplicate the symbology of the MoransICluster layer. Use the field LMiZScore Fixed 900 for the value field. Close the layer properties and turn on the Alarm Territories layer.

Compare the new results with the results of the first Cluster/Outlier analysis. The hot spots seem to correspond to portions of the major freeways, with a particular intersection in District 5 showing up as very dangerous. Perhaps freeways are one of the underlying causes of the hot spots.

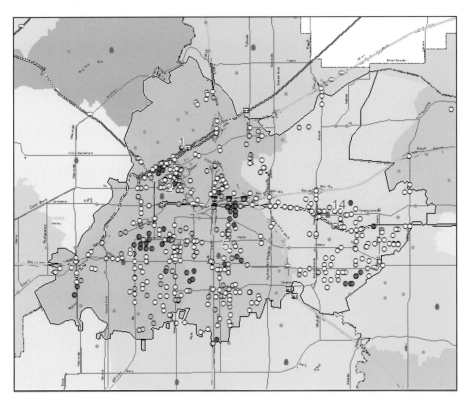

6 To determine which of the high z-scores were created by high values being close to other high values and which z-scores were the results of low values being close to other low values, set up a copy of layer MoransICluster900 and symbolize the COType field as you did with the ClusterTypes layer earlier in this tutorial.

- Copy the MoransICluster900 layer.
- Name the copied layer **ClusterTypes900**.
- Change the symbology type to Unique Values.
- Set the value field to COType.
- Add the values HH, HL, LH, and LL, and clear all other values.
- Change the symbol to Circle 1 with a size of 50.
- Change circle colors.
- Set the transparency to 40%.
- Move the layer below MoransICluster900 in the table of contents.

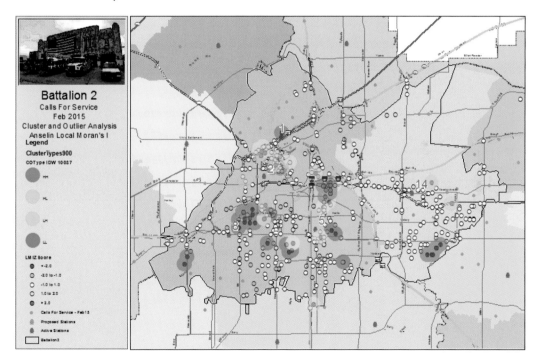

7 Save your map document as **Tutorial 9-1.mxd** to the \GIST2\MyExercises folder. If you are not continuing to the exercise, exit ArcMap.

Exercise 9-1

The tutorial showed how to create a feature class to demonstrate the results of the Cluster and Outlier Analysis tool. From this analysis, you can compare the data with other datasets to search for a correlation.

In this exercise, you will repeat the process using different datasets. The Incident_Jan15 data can also be used for this type of analysis, using the FEE values for the weight.

- Open Tutorial 9-1E.mxd.
- Run the Cluster and Outlier Analysis tool. Use FEE values as the weight.
- Increase the symbol size of the output, and symbolize the CensusBlkGrpIncome layer with graduated colors using the Median Household Income field (determined by the census metadata).
- Create symbology showing cluster types.
- Add a text box discussing any possible relationships between the calls for service and median household income.
- Change the titles, colors, and legend to make a visually pleasing map.
- Repeat the analysis using the Cluster and Outlier Analysis tool with a fixed-distance band of 900 feet to reduce the threshold errors.
- Create the symbology showing the cluster types for this new layer as well.
- Save the results as **Exercise 9-1.mxd** to the \GIST2\MyExercises folder.

What to turn in

If you are working in a classroom setting with an instructor, you may be required to submit the maps you created in tutorial 9-1.

Turn in a printed map or screen capture of the following:

       Tutorial 9-1.mxd (two maps)
       Exercise 9-1.mxd (two maps)

Tutorial 9-1 review

The Cluster and Outlier Analysis tool ran the pattern-analysis calculations and symbolized the results. The Moran's I index was calculated to show the level of clustering, with the positive values showing clustering and the negative values showing dispersion. The z-scores were also calculated for the features to help you decide if the clustering is statistically significant. High z-scores may result from features with high values being surrounded by other features with high values, or from features with low values being surrounded by other features with low values. Features with low values surrounded by features with high values—or conversely, features with high values surrounded by features with low values—produce a low z-score. The field COType is used to determine the type of clustering observed.

The results were also displayed over another dataset to try to visually determine if any underlying factors exist. The census data shows total population, and is compared to the hot spots. No association seems to exist, so other factors should be investigated. The exercise includes income levels and provides another possible factor in the clustering.

Study questions

1. What relationship does the z-score have to the Moran's I index?

2. What other data might be having an effect on the calls for service?

3. One analysis uses the inverse distance as the conceptualization of spatial relationships, and the other uses a fixed-distance band. What is the difference between these two settings?

Other real-world examples

An oil well drilling company may look at the clustering of high output values to site new wells. Areas with clustering of low values might be avoided.

A testing agency might look for clustering of high school test scores within the school district to find schools that excel, or schools that do poorly, to see if there is an association between the area and the scores.

9-1

9-2

Tutorial 9-2

Performing hot spot analysis

The Getis-Ord hot spot analysis shows where high and low values are clustered. The tool compares the values of each feature with the neighboring features within a user-specified distance. The values for each feature are then color-coded to show high- and low-value clusters.

Learning objectives

- *Evaluate z-scores*
- *Analyze high- and low-value clustering*
- *Classify symbols*

Preparation

- *Read pages 175–90 in* The Esri Guide to GIS Analysis, *volume 2.*

Introduction

The Gi statistic, also called Gi* (pronounced *G-i-star*), uses both the location and the value in the pattern calculations. The results of the Gi* tool are used to see the effect of the value field on the clustering over the user-specified distance. The distance is determined by the characteristics of the input dataset. Features that represent large, wide groupings may use a larger distance band value, while features that represent a local region or smaller feature areas might use a small distance band value. With a larger value, expect to get a few large clusters. A smaller distance value may result in many, smaller clusters.

Once the calculations are completed, the results are symbolized to display the clustering of high and low values. The Hot Spot Analysis tool with Rendering sets the classification for the new layer with the low-value clusters shown in dark blue and the high-value clusters shown in red.

Scenario The Tarrant County Economic Development Office has gotten word of the great maps you made for the fire department and wants to get in on the act. The board wants to see where the median income per household clusters. It wants to direct charity-collection efforts to areas with the largest clusters of high income, and target job creation in areas with the lowest median income. The charity efforts will help support many county initiatives, and the job opportunities may help raise the standard of living for struggling families.

You will create a map that shows the areas in which the median income values cluster using the Gi* statistic.

Data   The first set of data is the census block-group-level data. It contains a field for the median income, P053001, which can be used in the Gi* function. The area you need is already cut out for use, but the analysis can be repeated for any area using the exercise data.

The second set of data is the street network data to give the map some reference. It comes from the Data and Maps for ArcGIS data.

Map high-income and low-income clusters

1   In ArcMap, open Tutorial 9-2.mxd.

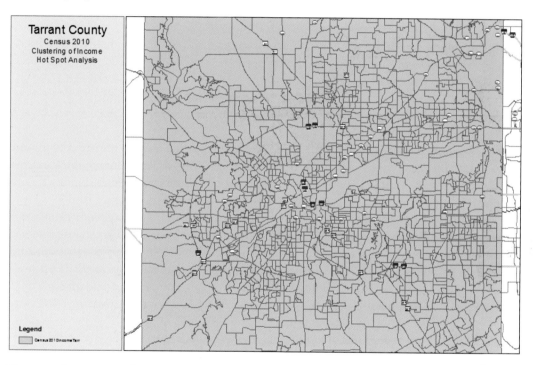

The map shows a simple setup. The county roads are shown with the 2010 Census data, which includes income. The Gi* function can be run on the Census2010IncomeTarr layer using the P053001 field as the input value.

2   Use the Search window to locate and open the Hot Spot Analysis with Rendering tool.

3   Enter the parameters as follows:

- Input Feature Class: Census2010IncomeTarr
- Input Field: P053001 (median income)
- Output Layer File: \GIST2\MyExercises\GetisTarrant.lyr
- Output Feature Class: \GIST2\MyExercises\MyData.gdb\GetisTarrant
- Distance Band: **5280** feet (one mile)

4 When your dialog box matches the graphic, click OK.

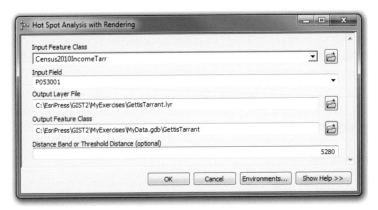

A new feature class is created, and added to the table of contents. The Gi* statistic automatically symbolizes the results.

5 Turn off the Census2000IncomeTarr layer, and make sure the GetisTarrant layer is visible.

The dark-blue areas represent a clustering of low values, and the red areas represent a clustering of high values. This dramatic result gives a clear indication of where to concentrate the job-creation programs and charity-collection efforts.

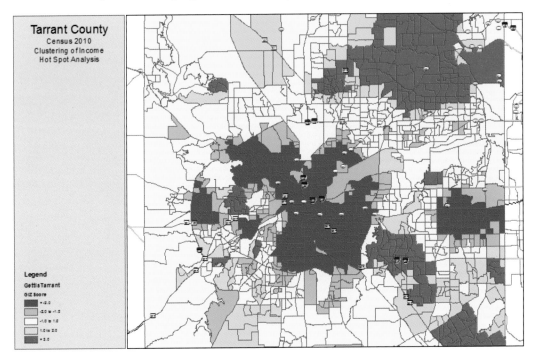

6 Save your map document as **Tutorial 9-2.mxd** to the \GIST2\MyExercises folder. If you are not continuing to the exercise, exit ArcMap.

Exercise 9-2

The tutorial showed how to create a feature class to demonstrate the results of the Gi* Analysis tool. The results were automatically symbolized to show high- and low-value clustering.

In this exercise, you will repeat the process using different datasets. The Dallas County data is provided to perform the same Gi* analysis, using the field P053001 for the value field.

- Open Tutorial 9-2E.mxd.
- Run the Hot Spot Analysis (Getis-Ord Gi*) tool. Use the P053001 field as the value.
- Set the Distance Band value to 5280 feet.
- Add a text box with a note describing the significance of the results.
- Change the titles, colors, and legend to make a visually pleasing map.
- Save the results as **Exercise 9-2.mxd** to the \GIST2\MyExercises folder.

What to turn in

If you are working in a classroom setting with an instructor, you may be required to submit the maps you created in tutorial 9-2.

Turn in a printed map or screen capture of the following:
    Tutorial 9-2.mxd
    Exercise 9-2.mxd

Tutorial 9-2 review

The Gi* Hot Spot Analysis tool is simple in its execution, and dramatic in its results. The median household income is used as the input value, along with the distance band value, to determine how one feature compares to its neighbors. Areas in which high values cluster are symbolized in red, and areas in which low values cluster are shown in blue.

This clarity of results is due to the choice of an optimal distance band value. The high value results in a few large areas of clustering. This analysis will help the county plan its efforts more effectively.

Study questions

1. What results might you expect with a smaller distance band value?

2. How does the Gi* statistic differ from the Moran's I clustering tool?

Other real-world examples

Police departments use the Gi* statistic to identify hot spots of crime by severity or frequency. The crimes can be ranked with the highest numbers assigned to felonies and the lowest assigned to misdemeanors. The Getis-Ord Gi* analysis will show where the higher ranks cluster, and where the lower ranks cluster. Crimes may also be aggregated to regions such as city blocks and analyzed using the Getis-Ord Gi* tool to show the hot spots or cold spots of crime frequency.

The city housing authority may geocode the addresses of reported foreclosures within the city limit. It can aggregate these locations by census block and then analyze them using the Getis-Ord Gi* tool. The hot spots for foreclosures can be overlaid on various demographic information to investigate underlying causes.

Independent projects

The book's nine chapters, with 70 tutorials and exercises, demonstrate many analysis techniques. These techniques range from simple visual analysis to complex spatial statistics. Now it is your turn to demonstrate mastery of these concepts.

The beginning of each chapter includes a list of learning objectives. These objectives are tasks that everyone who completes the book should be able to perform.

This final section can be used as a class project or for individuals to challenge themselves. It involves taking the exercise data provided with this book, available in the Esri Press online resources at esri.com/esripress-resources, and developing an analysis project from scratch. The steps should include the following:
- Present the idea.
- Outline the process.
- Validate the data.
- Perform the analysis.
- Validate the outcome.
- Present the results.

Choose one of the following scenarios or write one of your own. Then perform all the steps for the analysis project as listed here. All the scenarios use Fort Worth Fire Department data for Battalion 2. As a bonus, include some of the spatial statistics methods in your analysis project.

Scenario 1

As a quick visual analysis, the fire chief wants to see each month's call-for-service data categorized by call severity. Use the FEE field, with the highest number being the most severe. Overlay a one-mile ring around all the stations in Battalion 2 and summarize the calls by severity for each station. Include on the map a table that shows the results. Make a map series or animation of March 2006 to February 2007.

Scenario 2

The fire chief utters one simple question that is actually a complex analysis project: "How many calls did we run in Battalion 2 that were out of a station's regular service area?"

In the incident data, the field "station" contains the number of the station that responded. Use the polygon boundary of service areas to answer this question. Make a map series or animation of March 2006 to February 2007.

Scenario 3

The fire chief again tosses out a complex question: "What is the average drive time to a call for each station?"

Use the Tarrant Roads centerline file to make a network database. Then use the Network Analyst tools to map the route from each station to every call it made. Get the average time in minutes for each route, assuming an average driving speed of 40 miles per hour (mph). Make a map series or animation of March 2006 to February 2007.

Scenario 4

As a review for the city's insurance group, make a map that shows drive time along a network from each of the stations in Battalion 2. The times should be one, two, and three minutes using an average speed of 40 mph. Then summarize how many responses were made within each of these distances, and how many were made outside these distances. The summaries can be totals for Battalion 2, including all its stations—for instance, all the calls within one minute of a station, within two minutes of a station, within three minutes of a station, and outside three minutes of a station. Make a map series or animation of March 2006 to February 2007.

Scenario 5

Demonstrate how the mean weighted center of calls for service for each station moves throughout the year. Show the movement of the weighted standard deviational ellipse for each station as well. Make a map series or animation of March 2006 to February 2007.

Scenario 6

Determine a suitable z-score and perform a hot spot analysis for each month's incident data. Make a map series or animation of March 2006 to February 2007.

Reference materials

- Class notes
- ArcGIS for Desktop Help
- *The Esri Guide to GIS Analysis*, volumes 1 and 2
- Tutorials and exercises in this book

Appendix A

Task index

Software tool/concept, **tutorial(s) in which it appears**

Data source credits

Chapter 1 data sources include:

\\EsriPress\GIST2\Data\City Of Oleander.gdb\LandRecords\LotBoundaries, derived from City of Euless.

\\EsriPress\GIST2\Data\City Of Oleander.gdb\LandRecords\Parcels, derived from City of Euless.

\\EsriPress\GIST2\Data\City Of Oleander.gdb\LandRecords\Zoning, derived from City of Euless.

Chapter 2 data sources include:

\\EsriPress\GIST2\Data\Census.gdb\DFWRegion\CensusBlkGrp, from Data and Maps for ArcGIS (2005), courtesy of TANA/GDT, US Census, Esri BIS (Pop2004).

\\EsriPress\GIST2\Data\Census.gdb\DFWRegion\CensusBlkGrpIncome, from Data and Maps for ArcGIS (2005), courtesy of TANA/GDT, US Census, Esri BIS (Pop2004).

\\EsriPress\GIST2\Data\Census.gdb\DFWRegion\CensusBlocks, from Data and Maps for ArcGIS (2005), courtesy of TANA/GDT, US Census, Esri BIS (Pop2004).

\\EsriPress\GIST2\Data\Census.gdb\MajorRoads, from Data and Maps for ArcGIS (2005), courtesy of TANA/GDT.

\\EsriPress\GIST2\Data\City Of Oleander.gdb\LandRecords\LotBoundaries, derived from City of Euless.

\\EsriPress\GIST2\Data\City Of Oleander.gdb\LandRecords\Parcels, derived from City of Euless.

\\EsriPress\GIST2\Data\FoodStoresHispanic.csv, created by author.

\\EsriPress\GIST2\Data\FoodStoresHispanic.xls, created by author.

\\EsriPress\GIST2\Data\FoodStoresPlus.csv, created by author.

\\EsriPress\GIST2\Data\FoodStoresPlus.xls, created by author.

Chapter 3 data sources include:

\\EsriPress\GIST2\Data\Census.gdb\DFWRegion\CensusBlkGrp, from Data and Maps for ArcGIS (2005), courtesy of TANA/GDT, US Census, Esri BIS (Pop2004).

\\EsriPress\GIST2\Data\Census.gdb\MajorRoads, from Data and Maps for ArcGIS (2005), courtesy of TANA/GDT.

\\EsriPress\GIST2\Data\City Of Oleander.gdb\LandRecords\City_Limit, derived from City of Euless.

\\EsriPress\GIST2\Data\City Of Oleander.gdb\TreeInventory, derived from City of Euless.

Chapter 4 data sources include:

\\EsriPress\GIST2\Data\City Of Oleander.gdb\FloodPlain\FloodAreas, derived from City of Euless.

\\EsriPress\GIST2\Data\City Of Oleander.gdb\FloodPlain\FloodZone, derived from City of Euless.

\\EsriPress\GIST2\Data\City Of Oleander.gdb\LandRecords\LotBoundaries, derived from
City of Euless.

\\EsriPress\GIST2\Data\City Of Oleander.gdb\LandRecords\Parcels, derived from City of Euless.

Chapter 5 data sources include:

\\EsriPress\GIST2\Data\Census.gdb\MajorRoads, from Data and Maps for ArcGIS (2005), courtesy of
TANA/GDT.

\\EsriPress\GIST2\Data\Census.gdb\TarrantCounty\TarrantLakes, derived from North Central Texas
Council of Governments.

\\EsriPress\GIST2\Data\Census.gdb\TarrantCounty\TarrantRivers, derived from North Central Texas
Council of Governments.

\\EsriPress\GIST2\Data\Census.gdb\TarrantCounty\TarrantRoads, derived from North Central Texas
Council of Governments.

\\EsriPress\GIST2\Data\City Of Oleander.gdb\LandRecords\LotBoundaries, derived from City of
Euless.

\\EsriPress\GIST2\Data\City Of Oleander.gdb\LandRecords\Parcels, derived from City of Euless.

\\EsriPress\GIST2\Data\City Of Oleander.gdb\DFWCities, derived from North Central Texas Council of
Governments.

\\EsriPress\GIST2\Data\City Of Oleander.gdb\FireDepartment\AmbulanceRuns0514, derived from
City of Euless.

\\EsriPress\GIST2\Data\City Of Oleander.gdb\FireDepartment\FireDistances, derived from City of
Euless.

\\EsriPress\GIST2\Data\City Of Oleander.gdb\FireDepartment\FireRuns0514, derived from
City of Euless.

\\EsriPress\GIST2\Data\City Of Oleander.gdb\FireDepartment\Stations, derived from City of Euless.

\\EsriPress\GIST2\Data\City Of Oleander.gdb\LandRecords\Lot_Buffer, created by author.

\\EsriPress\GIST2\Data\City Of Oleander.gdb\Planimetric_Data\BuildingFootprints, derived from
City of Euless.

\\EsriPress\GIST2\Data\City Of Oleander.gdb\Planimetric_Data\Creeks, derived from City of Euless.

\\EsriPress\GIST2\Data\City Of Oleander.gdb\Planimetric_Data\Lakes, derived from City of Euless.

\\EsriPress\GIST2\Data\City Of Oleander.gdb\CityLimit, derived from City of Euless.

\\EsriPress\GIST2\Data\City Of Oleander.gdb\WestNile, derived from City of Euless.

\\EsriPress\GIST2\Data\City Of Oleander.gdb\WestNileStudyArea, derived from City of Euless.

\\EsriPress\GIST2\Data\Networks.gdb\NetworkAnalysis\FireStations, derived from City of Euless.

\\EsriPress\GIST2\Data\Networks.gdb\NetworkAnalysis\StreetCenterlines, derived from City of Euless.

\\EsriPress\GIST2\Data\Planimetric.gdb\Data_2015\BodiesOfWater, derived from City of Euless.

\\EsriPress\GIST2\Data\Planimetric.gdb\Data_2015\Creeks, derived from City of Euless.

\\EsriPress\GIST2\Data\Planimetric.gdb\Data_2015\MiscStructures, derived from City of Euless.

\\EsriPress\GIST2\Data\Planimetric.gdb\Data_2015\Recreation, derived from City of Euless.

\\EsriPress\GIST2\Data\Planimetric.gdb\Data_2015\Street_Centerlines, derived from City of Euless.

\\EsriPress\GIST2\Data\Planimetric.gdb\Data_2015\Trails, derived from City of Euless.

\\EsriPress\GIST2\Data\Planimetric.gdb\Data_2015\TreeInventory, derived from City of Euless.

\\EsriPress\GIST2\Data\Planimetric.gdb\Data_2015\TreeInventory_Midway, derived from City of Euless.

\\EsriPress\GIST2\Data\Planimetric.gdb\Data_2015\TreeInventory_Oleander, derived from City of
Euless.

\\EsriPress\GIST2\Data\AmbulanceRuns2014.csv, derived from City of Euless.

Chapter 6 data sources include:

\\EsriPress\GIST2\Data\Census.gdb\MajorRoads, from Data and Maps for ArcGIS (2005), courtesy of TANA/GDT.

\\EsriPress\GIST2\Data\City Of Oleander.gdb\FireDepartment\Level1, created by author.

\\EsriPress\GIST2\Data\City Of Oleander.gdb\FireDepartment\Level2, created by author.

\\EsriPress\GIST2\Data\City Of Oleander.gdb\FireDepartment\Level3, created by author.

\\EsriPress\GIST2\Data\City Of Oleander.gdb\FireDepartment\Level4, created by author.

\\EsriPress\GIST2\Data\City Of Oleander.gdb\FireDepartment\Site1_Time, created by author.

\\EsriPress\GIST2\Data\City Of Oleander.gdb\FireDepartment\Site2_Level1, created by author.

\\EsriPress\GIST2\Data\City Of Oleander.gdb\FireDepartment\Site2_Level2, created by author.

\\EsriPress\GIST2\Data\City Of Oleander.gdb\FireDepartment\Site2_Level3, created by author.

\\EsriPress\GIST2\Data\City Of Oleander.gdb\FireDepartment\Site2_Time, created by author.

\\EsriPress\GIST2\Data\City Of Oleander.gdb\FireDepartment\TornadoApril07, created by author.

\\EsriPress\GIST2\Data\City Of Oleander.gdb\FireDepartment\TornadoDrill, created by author.

\\EsriPress\GIST2\Data\City Of Oleander.gdb\FireDepartment\TornadoPath_Buffer, created by author.

\\EsriPress\GIST2\Data\City Of Oleander.gdb\FireDepartment\TornadoPathApril07, created by author.

\\EsriPress\GIST2\Data\City Of Oleander.gdb\FireDepartment\TornadoPathDrill, created by author.

\\EsriPress\GIST2\Data\City Of Oleander.gdb\LandRecords\LotBoundaries, derived from City of Euless.

\\EsriPress\GIST2\Data\City Of Oleander.gdb\LandRecords\ParcelTaxValue, derived from City of Euless.

\\EsriPress\GIST2\Data\City Of Oleander.gdb\Planimetric_Data\BuildingFootprints, derived from City of Euless.

Chapter 7 data sources include:

\\EsriPress\GIST2\Data\Census.gdb\DFWRegion\CensusBlkGrp, from Data and Maps for ArcGIS (2005), courtesy of TANA/GDT, US Census, Esri BIS (Pop2004).

\\EsriPress\GIST2\Data\Census.gdb\MajorRoads, from Data and Maps for ArcGIS (2005), courtesy of TANA/GDT.

\\EsriPress\GIST2\Data\City of Ft Worth.gdb\Fire_Department\Active_Stations, derived from City of Fort Worth.

\\EsriPress\GIST2\Data\City of Ft Worth.gdb\Fire_Department\Alarm_Territories, derived from City of Fort Worth.

\\EsriPress\GIST2\Data\City of Ft Worth.gdb\Fire_Department\Battalions, derived from City of Fort Worth.

\\EsriPress\GIST2\Data\City of Ft Worth.gdb\Fire_Department\Incident_Feb15, derived from City of Fort Worth.

\\EsriPress\GIST2\Data\City of Ft Worth.gdb\Fire_Department\Proposed_Stations, derived from City of Fort Worth.

\\EsriPress\GIST2\Data\City of Ft Worth.gdb\Fire_Department\Storm_Track, derived from City of Fort Worth.

\\EsriPress\GIST2\Data\City of Ft Worth.gdb\Fire_Department\Storm_Track_Dallas, derived from City of Fort Worth.

\\EsriPress\GIST2\Data\City of Ft Worth.gdb\Fire_Department\Tornado, derived from City of Fort Worth.

\\EsriPress\GIST2\Data\City of Ft Worth.gdb\Fire_Department\Tornado_Dallas, derived from City of Fort Worth.

\\EsriPress\GIST2\Data\City Of Oleander.gdb\LandRecords\LotBoundaries, derived from
City of Euless.

\\EsriPress\GIST2\Data\City Of Oleander.gdb\LandRecords\Parcels, derived from City of Euless.

\\EsriPress\GIST2\Data\Library.gdb\Districts, created by author.

\\EsriPress\GIST2\Data\Library.gdb\Grid300, created by author.

\\EsriPress\GIST2\Data\Library.gdb\PatronLocations, derived from City of Euless.

Chapter 8 data sources include:

\\EsriPress\GIST2\Data\Census.gdb\MajorRoads, from Data and Maps for ArcGIS (2005), courtesy of
TANA/GDT.

\\EsriPress\GIST2\Data\Census.gdb\TarrBlkGrp, from Data and Maps for ArcGIS (2005), courtesy of
TANA/GDT.

\\EsriPress\GIST2\Data\City of Ft Worth.gdb\Fire_Department\Active_Stations, derived from City of
Fort Worth.

\\EsriPress\GIST2\Data\City of Ft Worth.gdb\Fire_Department\Alarm_Territories, derived from City
of Fort Worth.

\\EsriPress\GIST2\Data\City of Ft Worth.gdb\Fire_Department\Battalion2, derived from City of
Fort Worth.

\\EsriPress\GIST2\Data\City of Ft Worth.gdb\Fire_Department\Battalions, derived from City of
Fort Worth.

\\EsriPress\GIST2\Data\City of Ft Worth.gdb\Fire_Department\Incident_Feb15, derived from City of
Fort Worth.

\\EsriPress\GIST2\Data\City of Ft Worth.gdb\Fire_Department\Incident_Jan15, derived from City of
Fort Worth.

\\EsriPress\GIST2\Data\City of Ft Worth.gdb\Fire_Department\Grid200, derived from City of
Fort Worth.

\\EsriPress\GIST2\Data\City of Ft Worth.gdb\Fire_Department\Grid500, derived from City of
Fort Worth.

\\EsriPress\GIST2\Data\City of Ft Worth.gdb\Fire_Department\Proposed_Stations, derived from
City of Fort Worth.

\\EsriPress\GIST2\Data\City of Ft Worth.gdb\Fire_Department\TarrantStreets, derived from North
Central Texas Council of Governments.

\\EsriPress\GIST2\Data\AmbulanceRuns2014.csv, derived from City of Fort Worth.

\\EsriPress\GIST2\Data\CodeFee.dbf, derived from City of Fort Worth.

\\EsriPress\GIST2\Data\FireRuns2014.xlsx, derived from City of Fort Worth.

\\EsriPress\GIST2\Data\FireRuns2014.csv, derived from City of Fort Worth.

Chapter 9 data sources include:

\\EsriPress\GIST2\Data\Census.gdb\MajorRoads, from Data and Maps for ArcGIS (2005), courtesy of
TANA/GDT.

\\EsriPress\GIST2\Data\Census.gdb\TarrantCounty\CensusEthnic, derived from North Central Texas
Council of Governments.

\\EsriPress\GIST2\Data\City of Ft Worth.gdb\Fire_Department\Active_Stations, derived from City of
Fort Worth.

\\EsriPress\GIST2\Data\City of Ft Worth.gdb\Fire_Department\Alarm_Territories, derived from
City of Fort Worth.

\\EsriPress\GIST2\Data\City of Ft Worth.gdb\Fire_Department\Battalion2, derived from City of Fort Worth.

\\EsriPress\GIST2\Data\City of Ft Worth.gdb\Fire_Department\Battalions, derived from City of Fort Worth.

\\EsriPress\GIST2\Data\City of Ft Worth.gdb\Fire_Department\CensusIncome, from Data and Maps for ArcGIS (2005), courtesy of TANA/GDT.

\\EsriPress\GIST2\Data\City of Ft Worth.gdb\Fire_Department\CensusIncomeTarr, Data and Maps for ArcGIS (2005), courtesy of TANA/GDT, US Census, Esri BIS (Pop2004).

\\EsriPress\GIST2\Data\City of Ft Worth.gdb\Fire_Department\CensusIncomeDall, Data and Maps for ArcGIS (2005), courtesy of TANA/GDT, US Census, Esri BIS (Pop2004).

\\EsriPress\GIST2\Data\City of Ft Worth.gdb\Fire_Department\Incident_Feb15, derived from City of Fort Worth.

\\EsriPress\GIST2\Data\City of Ft Worth.gdb\Fire_Department\Incident_Jan15, derived from City of Fort Worth.

\\EsriPress\GIST2\Data\City of Ft Worth.gdb\Fire_Department\Proposed_Stations, derived from City of Fort Worth.

"Independent projects" data sources include:

\\EsriPress\GIST2\Data\City of Ft Worth.gdb\Fireort_Department\Incident_Apr14, derived from City of Fort Worth.

\\EsriPress\GIST2\Data\City of Ft Worth.gdb\Fire_Department\Incident_Aug14, derived from City of Fort Worth.

\\EsriPress\GIST2\Data\City of Ft Worth.gdb\Fire_Department\Incident_Dec14, derived from City of Fort Worth.

\\EsriPress\GIST2\Data\City of Ft Worth.gdb\Fire_Department\Incident_Feb15, derived from City of Fort Worth.

\\EsriPress\GIST2\Data\City of Ft Worth.gdb\Fire_Department\Incident_Jan15, derived from City of Fort Worth.

\\EsriPress\GIST2\Data\City of Ft Worth.gdb\Fire_Department\Incident_Jul14, derived from City of Fort Worth.

\\EsriPress\GIST2\Data\City of Ft Worth.gdb\Fire_Department\Incident_Jun14, derived from City of Fort Worth.

\\EsriPress\GIST2\Data\City of Ft Worth.gdb\Fire_Department\Incident_Mar14, derived from City of Fort Worth.

\\EsriPress\GIST2\Data\City of Ft Worth.gdb\Fire_Department\Incident_May14, derived from City of Fort Worth.

\\EsriPress\GIST2\Data\City of Ft Worth.gdb\Fire_Department\Incident_Nov14, derived from City of Fort Worth.

\\EsriPress\GIST2\Data\City of Ft Worth.gdb\Fire_Department\Incident_Oct14, derived from City of Fort Worth.

\\EsriPress\GIST2\Data\City of Ft Worth.gdb\Fire_Department\Incident_Sep14, derived from City of Fort Worth.

Data license agreement

Downloadable data that accompanies this book is covered by a license agreement that stipulates the terms of use.